Report 143

1995
Reprinted 2005

The Standard Penetration Test (SPT): Methods and Use

C.R.I. Clayton MSc DIC PhD CEng MICE CGeol FGS

CIRIA *sharing knowledge ■ building best practice*

Classic House, 174–180 Old Street, London EC1V 9BP
TELEPHONE 020 7549 3300 FAX 020 7253 0523
EMAIL enquiries@ciria.org
WEBSITE www.ciria.org

Summary

This report provides guidance on the use of the Standard Penetration Test (SPT) in geotechnical engineering. Used worldwide in a variety of ground conditions, this simple test is relied upon to provide information about the properties of soils and weak rocks and as a basis for design. From a review of the historical use of the SPT and an analysis of the testing procedures used in UK and abroad, the report examines the variability of SPT results, whether caused by the methods of drilling and testing or influenced by ground conditions. The principal uses of the SPT are examined, i.e. ground investigation profiling, soil classification, the determination of geotechnical design parameters and for direct design. Suggested methods are proposed and outlined for the determination from SPT results of geotechnical parameters for granular soils, cohesive coils, weak rocks and chalk. Similarly, suggested direct design methods are discussed for the settlements of shallow foundations on granular soil, for piles in soils, weak rocks and chalk, and for estimating liquefaction potential and sheet-pile driveability in granular soils.

In order that better use can be made of the SPT in the future, the report concludes with recommendations covering field practice, interpretation of results and their application.

In addition to a substantial reference list, there are two appendices which present the International Reference Test Procedure for the SPT and an explanation of SPT energy measurement.

The Standard Penetration Test (SPT): methods and use
Construction Industry Research and Information Association
CIRIA Report 143, 1995

ISBN 0 86017 4190
ISSN 0305 408X
© CIRIA 1995
Reprinted 2005

Keywords	Reader interest	Classification	
Geotechnics, *in-situ* testing, penetration testing, SPT, foundations, pile design, SPT energy measurement	Geotechnical engineers, structural engineers	AVAILABILITY	Unrestricted
		CONTENT	Subject area review
		STATUS	Committee guided
		USER	Geotechnical and structural engineers

This book is a reprint of the edition published in 1995 and reflects best practice at that time. It also contains addresses and telephone numbers for a large number of organisations which have not been updated. The names, locations and structure of many of these organisations will have changed.

Additionally, government reorganisation has meant that DETR responsibilities have been moved variously to the Department of Trade and Industry (DTI), the office of the Deputy Prime Minister (ODPM), and the Department for Transport. References made to government agencies in this publication should be read in this context. For clarification, readers should contact the Department of Trade and Industry.

Published by CIRIA, Classic House, 174–180 Old Street, London, EC1V 9BP. All rights reserved. No part of this publication may be reproduced or transmitted in any form or by any means, including photocopying and recording, without the written permission of the copyright holder, application for which should be addressed to the publisher. Such written permission must also be obtained before any part of this publication is stored in a retrieval system of any nature.

Foreword

This report presents the results of a research project in CIRIA's ground engineering programme on *in-situ* testing. The report was written by Professor C.R.I. Clayton of the University of Surrey under contract to CIRIA.

Following CIRIA's usual practice, the research was guided by a Steering Group which comprised:

Mr J.M. Head	now Frank Graham Consulting Engineers Ltd
Mr D.J. Mallard	Nuclear Electric plc
Mr I.K. Nixon	Consultant
Dr M.A. Stroud	Ove Arup and Partners
Mr S. Thorburn	Thorburn Group
Mr K.W. Vickery	Cementation Piling and Foundations Ltd

CIRIA's research managers for this project were Mr C P Wynne, initially, and Mr F M Jardine.

ACKNOWLEDGEMENTS

The project was funded by the Construction Directorate, Department of the Environment, and CIRIA.

CIRIA and the author are grateful for the help given to this project by the funders, the members of the Steering Group, and by the many individuals who were consulted. The author particularly acknowledges the help with particular aspects of the report given by A. Brown, Professor L. Decourt, Professor V.F.B. de Mello, Dr D.A. Greenwood, K. Hawkins, Professor J.C. Hiedra Lopez, Dr J-C. Hiedra Cobo, B. A. Leach, Professor J. Milititsky, Dr O. Moretta, R.J. Newman, D. Norbury and Professor J.H. Schmertmann.

Contents

List of figures 6

List of tables 9

Notations 10

1   **INTRODUCTION** 13

2   **THE EVOLUTION OF THE STANDARD PENETRATION TEST (SPT)** 16

3   **UK PRACTICE** 20
    3.1   The British standard method of test 20
    3.2   UK boring techniques 22
    3.3   UK standard penetration test equipment 24
    3.4   Test procedures in current use 27
    3.5   Synopsis 28

4   **PRACTICE OUTSIDE THE UK** 29
    4.1   National standards 29
    4.2   The issmfe European and international reference test procedures .. 31
    4.3   International practice and equipment 33
    4.4   Further developments 40
    4.5   Synopsis 40

5   **THE INFLUENCE OF DIFFERENT PRACTICES AND EQUIPMENT ON PENETRATION RESISTANCE** 42
    5.1   Drilling or boring technique 43
    5.2   Test equipment 50
    5.3   Test method 55
    5.4   Synopsis 57

6   **THE INFLUENCE OF GROUND CONDITIONS ON PENETRATION RESISTANCE** 58
    6.1   The SPT in granular soil 59
    6.2   The SPT in cohesive soil 66
    6.3   Weak and weathered rocks 67
    6.4   Synopsis 70

7   **SPT APPLICATIONS** 71
    7.1   Profiling 73
    7.2   Classification 74
    7.3   Parameter determination 75
    7.4   Direct design methods 76

8   **DETERMINATION OF GEOTECHNICAL PARAMETERS** 77
    8.1   Estimation of parameters in granular soils 78
    8.2   Estimation of parameters in cohesive soils 84
    8.3   Estimation of parameters in weak rock 89
    8.4   Estimation of parameters in chalk 91

9 DIRECT DESIGN METHODS ... 95
9.1 Estimation of settlements of shallow foundations on granular soil ... 95
9.2 Design of piles in soils, weak rocks and chalk ... 101
9.3 Liquefaction potential in granular soils ... 105
9.4 Estimation of sheet pile drivability ... 107

10 RECOMMENDATIONS FOR THE USE OF THE SPT ... 109
10.1 SPT field methods ... 109
10.2 Interpretation ... 113
10.3 Supervision and training ... 115
10.4 Application of SPT test results ... 115
10.5 Further work ... 116

References ... 117

Appendix 1 International Reference Test Procedure IRTP ... 127

Appendix 2 SPT energy measurement ... 132

Figures

Figure 1	*Early SPT split-spoon sampler*	13
Figure 2	*Early SPT trip hammer*	14
Figure 3	*Manual lifting of the SPT weight on a washboring rig in Brazil in 1987*	15
Figure 4	*BS 1377 split-barrel samplers (a) BS 1377: 1975 (b) BS 1377: 1990*	18
Figure 5	*British Standard requirements for the SPT*	21
Figure 6	*Light percussion boring equipment*	22
Figure 7	*Automatic trip hammer in use*	25
Figure 8	*British automatic trip hammer*	26
Figure 9	*ASTM split-barrel sampler (ASTM D1586-67, re-approved 1974)*	29
Figure 10	*Automatic trip hammers (a) Israeli, (b) Japanese*	36
Figure 11	*Japanese hand-controlled SPT trip hammer components*	37
Figure 12	*US SPT hammers (a) Safety hammer (b) Donut hammer*	38
Figure 13	*Using a cathead to lift the SPT weight in Venezuela in 1987*	38
Figure 14	*Summary of effects of different practices and equipment on SPT N value*	43
Figure 15	*Results of four investigations in the Trent river gravels by light percussion boring*	44
Figure 16	*SPT results from investigations in the Chalk at Littlebrook*	45

Figure 17	*Comparison of SPT N values for different drilling methods for a site in the Chalk at Leatherhead.*	46
Figure 18	*Comparison of SPT(C) results obtained by two different ground investigation companies, using similar light percussion equipment at the same site in chalk.*	47
Figure 19	*Influence of borehole diameter and drilling method at Cairo*	48
Figure 20	*Comparison between SPT results in Norwich Crag sands, Suffolk, from light percussion and rotary borings,* *(a) with shell and casing* *(b) uncased, bentonite mud supported small diameter rotary borings*	49
Figure 21	*Influence of rod diameter on penetration resistance.*	52
Figure 22	*Effect of long rod lengths on transmitted energy*	53
Figure 23	*Effect of short rods on energy transmission and c_u/N ratio*	54
Figure 24	*Comparison of penetration resistance with the standard SPT shoe (SPT)*	54
Figure 25	*Comparison of SPT (i.e. open shoe) and SPT(C) (i.e. 60° cone) N values in chalk.*	55
Figure 26	*Comparison of SPT and SPT(C) N values for a site in the Chalk at Brighton.*	55
Figure 27	*Effect of omission of split-soon liners on penetration resistance*	56
Figure 28	*Corrections for overburden pressure and overconsolidation* *(a) correction factor for overburden pressure* *(b) correction factor for increase in mean stress due to overconsolidation*	62
Figure 29	*Effect of decrease of vertical effective stress, due to excavation, on SPT N values*	64
Figure 30	*Effect of grain size on SPT values,* *(a) effect of mean grain size on N_S/N_L ratio* *(b) tentative grain size correction for SPT*	65
Figure 31	*Correlation between N value and undrained shear strength*	66
Figure 32	*Correlation between SPT N value and average saturated moisture content*	68
Figure 33	*Correlations between SPT N value and chalk visual weathering grade.*	68
Figure 34	*Lack of correlation between SPT N value and visual for weathering grade in the Chalk of Hampshire.*	69
Figure 35	*Relationships between torque, T, and SPT value, N_{70} for* *(a) sandy alluvial soils and* *(b) residual soils derived from migmatites, granites and gneiss.*	76
Figure 36	*Suggested Relationships between* *(a) N, ϕ' and bearing capacity factors N_γ and N_q and* *(b) N, ϕ' and $\acute{o}_v'$.*	79

Figure 37	*Influence of relative density on the dilatancy component of effective angle of friction*	80
Figure 38	*Examples of correlations between equivalent drained and undrained modulus values and penetration resistance, N, for mudrocks, clays and granular soils*	82
Figure 39	*Relationship between stiffness, penetration resistance and degree of loading for sand*	83
Figure 40	*Comparison of the mean undrained shear strength-depth profiles in the London Clay determined from different test sizes*	85
Figure 41	*Correlation between coefficient f_2 (= N/m_v) and plasticity index*	87
Figure 42	*Relationship between effective modulus of overconsolidated clay, penetration resistance and degree of loading*	88
Figure 43	*Correlations between unconfined compressive strength and penetration resistance of weak rocks*	90
Figure 44	*Correlation of E' with SPT N value for chalk*	93
Figure 45	*Observed settlements of footings on sand of various densities*	96
Figure 46	*Schultze and Sherif's method (1973) for calculating the settlement of spread foundations on sand*	97
Figure 47	*Burland and Burbidge's (1985) method for estimating the settlement of granular soils*	99
Figure 48	*Maximum cyclic stress ratio versus penetration resistance for a magnitude 7.5 earthquake in clean sand*	108
Figure 49	*Effect of fines content on limiting cyclic stress ratio for sand*	108
Figure 50	*Effect of earthquake magnitude, M, on limiting cyclic stress ratio*	108
Figure 51	*Expected shear strain levels following liquefaction of dense sand*	108
Figure 52	*Cross section of SPT Sampler*	128
Figure 53	*Theoretical force-time relationship for a 63.5 kg hammer striking the end of an infinitely long AW rod*	136
Figure 54	*Layout of equipment during energy measurement*	139
Figure 55	*SPT energy correction for short rods and load cell position*	139
Figure 56	*SPT energy measuring systems*	140
Figure 57	*Energy measurement using the Oyo Geologger 3030 SPT measuring module*	142
Figure 58	*Typical force-time relationships from SPT blows*	143

Tables

Table 1	*National standards for the Standard Penetration Test*	30
Table 2	*SPT standards in different countries*	32
Table 3	*Approximate borehole diameter corrections proposed by Skempton (1986)*	47
Table 4	*Estimated influence of hammer type on SPT N value*	51
Table 5	*Comparison of in situ tests*	58
Table 6	*Influence of granular soil properties on dynamic penetration resistance*	60
Table 7	*Examples of direct and indirect applications of the SPT in geotechnical design*	72
Table 8	*SPT-based soil and rock classification systems*	74
Table 9	*Determination of parameters from SPT results*	77
Table 10	*Comparison of some correlations between G_{max} and N*	83
Table 11	*Young's modulus derived from Burland and Burbidge's I_c values*	84
Table 12	*Correlations between ultimate shaft resistance, f_s, of piles and SPT N value*	102
Table 13	*Estimation of ultimate shaft load for driven piles in chalk*	103
Table 14	*Correlations between end-bearing resistance, f_b, of piles and SPT N value*	104
Table 15	*Estimation of ultimate base load for piles in chalk*	105
Table 16	*Preliminary selection of sheet pile section to suit driving conditions in granular soils*	107
Table 17	*Examples of SPT records on rig foreman's daily reports*	113
Table 18	*Examples of SPT records on engineering report of borehole record*	114
Table 19	*Variation of rod energy ratios for SPT hammers*	133
Table 20	*SPT energy measurements on a DANDO automatic trip hammer*	143

Notation

A	cross-section area, gross cross-sectional area of SPT shoe
α	material dependent factor defined by Skempton (1986) relating relative density and penetration resistance
α	factor relating G_{max} and N
α_{max}	maximum acceleration at ground surface
β	foundation width
b	material dependent factor defined by Skempton (1986) relating relative density and penetration resistance
b	factor relating G_{max} and N
C_c	compression index
C_d	correction factor for borehole diameter
C_f	compaction factor (Hobbs and Healy, 1979)
C_n	overburden pressure correction factor for normally consolidated sands
C_{nk}	correction factor for increase in mean stress due to overconsolidation
C_{sg}	correction factor for grain size
C_{oc}	factor to allow for the increase in penetration resistance as a result of overconsolidation (Skempton, 1986)
c_u	undrained shear strength
$c_{u(100)}$	undrained shear strength measured on 100mm diameter specimens in the triaxial apparatus
D_r	relative density
D_{50}	mean particle size
d	shaft diameter of pile
d_s	depth of compressible stratum (Schultze and Sherif, 1973)
E	input energy
E	Young's modulus
E'	drained (effective) Young's modulus
E_i'	intact Young's modulus
E_r	energy delivered by SPT hammer per blow
E_u	undrained Young's Modulus
f	influence factor for geometry of foundation
f_1	factor relating undrained shear strength, c_u, to Standard Penetration Test resistance, N
f_1	depth factor (Burland and Burbidge, 1985)
f_2	factor relating coefficient of volume compressibility, m_v, to Standard Penetration Resistance, N
f_b	end-bearing resistance of a pile
f_s	ultimate shaft resistance of a pile
f_s	shape factor in Burland and Burbidge (1985) method
f_t	time factor in Burland and Burbidge (1985) method
G_{max}	shear modulus at very small strain
g	gravitational constant
H_s	thickness of compressible soil below foundation
I_c	compression index (Burland and Burbidge, 1985)
K	factor relating N-value to ultimate base resistance of a pile
K_o	ratio of horizontal to vertical effective stress *in situ*
K_{onc}	normally consolidated effective stress ratio
L	foundation length

l	drive length
M	earthquake magnitude
m_v	coefficient of volume compressibility
N	Standard Penetration Test resistance (blows/foot or blows/300mm)
$\tilde{N}$	mean SPT-value over a specified depth
N'	Standard Penetration Test resistance corrected when N is greater than 15 in saturated fine or silty sands
N_1	equivalent Standard Penetration Test resistance under a vertical effective stress of 100 kPa (or 1 tonf/ft² or 1 kgf/cm²)
N_s	blows/300mm using SPT in different sized soils
N_L	blows/300 mm using Large Penetration Test
N_q	bearing capacity factor
N_z	N-value at depth z
$N\gamma$	bearing capacity factor
$N_{40.5}$	SPT value when using 40.5mm dia. rods
N_{50}	SPT value when using 50mm dia. rods
N_{60}	equivalent Standard Penetration Test resistance corrected to 60% of the theoretical free-fall hammer energy
$(N_1)_{60}$	equivalent Standard Penetration Test resistance under a vertical effective stress of 100 kPa, and corrected to 60% of the theoretical free-fall hammer energy
N_{σ_v}	standard penetration resistance under a vertical effective stress of σ_v
p	applied bearing pressure
p'	mean effective stress at failure
p_z'	effective overburden pressure at depth z
Q_b	ultimate base load of a pile
Q_s	ultimate shaft load of a pile
q	applied bearing pressure
q'	average effective foundation pressure ($q-u$)
q_a	allowable bearing pressure
q_b	ultimate base resistance of a pile
q_s	shaft adhesion, ultimate shaft resistance of a pile
q_u	ultimate bearing capacity
q_u	end-bearing resistance
q_{net}	net bearing pressure
q_{ult}	ultimate bearing capacity
R_t	proportion of end-of-construction settlement taking place during each log cycle of time after 3 years
R_3	proportion of end-of-construction time-dependent settlement occurring in the 3-year period following construction
r_d	stress reduction factor
s	coefficient of settlement (Schultze and Sherif, 1973)
t	time (years) since end of construction
u	pore pressure
z	depth, or pile length
V	volume of displaced soil
Z_1	depth of influence below a foundation

α	constant depending upon soil and pile type
β	constant depending upon soil and pile type
γ_l	shear strain levels once liquefaction occurs
Δz	length of pile element whose mid point is at a depth z below the surface
δ	angle of friction between pile shaft and soil
δ'	effective angle of friction between soil/rock and a wall or pile
ν	Poisson's Ratio
ν_u	undrained Poisson's Ratio
ρ	settlement
ρ_{max}	maximum expected settlement
σ'	mean effective stress
σ_c	unconfined compressive strength
σ_h'	horizontal effective stress
σ_o	total overburden stress at depth under consideration
σ_o'	effective overburden stress at depth under consideration
σ_{vo}'	maximum vertical effective stress at foundation level (burland and Burbidge, 1985)
σ_v'	vertical effective stress
τ_{av}	average applied shear stress
ϕ	effective angle of friction
$\phi'cv$	effective angle of friction of soil shearing at constant volume

Note: The above list does not include the notation for Appendix 2 where symbols are defined as they are used.

1 Introduction

The Standard Penetration Test, or SPT, is the most widely used *in-situ* test, in the UK and throughout the world, as an indicator of the density and compressibility of granular soils. It is also commonly used to check the consistency of stiff or stony cohesive soils and weak rocks. Estimation of the liquefaction potential of saturated granular soils for earthquake design is often based on these tests. Available design methods for both shallow and deep foundations rely heavily on SPT results.

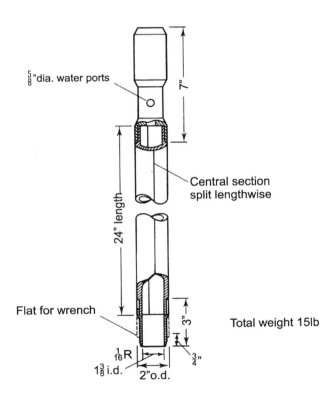

Figure 1 *Early SPT split-spoon sampler (after Terzaghi and Peck, 1948; Fletcher, 1965)*

The test consists of driving a standard 50-mm outside diameter thick-walled sampler into soil at the bottom of a borehole, using repeated blows of a 63.5-kg hammer falling through 760mm. The SPT N value is the number of blows required to achieve a penetration of 300mm, after an initial seating drive of 150mm. One of the advantages of the test is that it is carried out in routine exploration boreholes of varying diameters, so that (in contrast with other *in-situ* tests, such as the Cone Penetration Test or CPT) there is no need to bring special equipment to site.

Other advantages of the test are its simplicity, its low cost, and its ability to give a numerical parameter which may be related to crude but straightforward empirical design rules. For these reasons it is particularly valued in ground which is difficult to sample and test in the laboratory, such as sands, stony soils and weak rocks, and in variable ground. It has been, and continues to be, widely used despite the numerous valid criticisms of both the variability of procedures for making the test and the irrationality of many methods of interpreting its results.

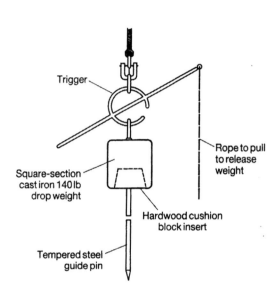

Figure 2   *Early SPT trip hammer (after Ireland et al., 1970)*

The purposes of this report are to bring together in one volume the many observations of causes of variability in SPT results, to recommend practices to achieve better workmanship and interpretation, and to guide foundation designers to a more rational use of SPT results.

Figure 3   *Manual lifting of the SPT weight on a washboring rig in Brazil in 1987*

2 The evolution of the Standard Penetration Test (SPT)

In the United States of America at the turn of the century, soil samples were generally wet and disturbed, being obtained by washboring. The earliest 'dry' samples are believed to have been taken by Colonel Charles R. Gow who, in 1902, developed a 1-in diameter sampler which was driven into the base of a borehole using a 110-lb hammer (Fletcher, 1965). In about 1927 two versions of a 2-in diameter 'split spoon' sampler were introduced, one by the Gow Co. (a subsidiary of the Raymond Concrete Pile Co.) and the other by Sprague and Henwood. The samplers were driven into the bottom of the borehole by repeated blows of a weight lifted by two men by hand, without the use of any powered-winch equipment (Mohr, 1937). Measurements of typical practice at this time showed that:

- the various hammers when weighed on a platform balance, had an average mass of about 140 lbs, and that
- the height to which two men could comfortably lift the weight with a normal swing of the arms was about 30 in (Mohr, 1966).

As part of the sampling record, the number of blows required to drive the sampler 12-in was recorded. These more or less standard sampling procedures first appeared in American specifications in the early 1930s. Figure 1 shows the early split-spoon sampler, after Terzaghi and Peck (1948), and Fletcher (1965). Figure 2, from Ireland et al. (1970), illustrates an early drive hammer. Figure 3 demonstrates the original method of hand lifting the weight.

Subsequent changes included the addition of a ball check-valve in the head of the sampler in an attempt to prevent sample loss and, from 1945, the use of A rods to connect the drive hammer at the top of the borehole to the split spoon sampler at the bottom of the borehole (Fletcher, 1965). The boreholes were generally relatively small diameter (up to 4-in) and were formed by washboring or augering techniques.

The earliest reference to a Standard Penetration Test procedure is in Terzaghi's 1947 paper to the Texas Conference on Soil Mechanics and Foundation Engineering. Thereafter, the SPT received worldwide attention in 1948, as a result of the publication of the first edition of Terzaghi and Peck's *Soil Mechanics in Engineering Practice*. The authors described a procedure whereby additional information on ground conditions could be obtained by measuring the driving resistance of the Raymond Concrete Pile Co.'s 2-in diameter soil sampling tube in a standard way, and commented that 'since this test furnishes vital information with very little extra cost it should never be omitted'. SPT results were used for a number of foundation design purposes, notably to determine the allowable bearing pressure of footings on sand. According to Bazaraa (1967), the database used by Terzaghi and Peck in 1948 included results from split spoons of different outside diameters (2-in and 1-5/8-in). Hvorslev (1949) makes no reference to a 'Standard Penetration Test' but gives details of four different dynamic penetration tests then in use in the USA, one of which is taken from Terzaghi and Peck (1948). It is therefore likely that Terzaghi coined the term 'Standard Penetration Test' in order to bring some uniformity to already diverse procedures.

The test procedure was not standardised in 1947, probably because the effects of using different procedures were not thought to be significant. Terzaghi (1947) describes the original procedure as follows:

> 'Before the penetration test is started the bottom of the hole is cleaned by means of a water jet or an auger. After the spoon reaches the bottom the drop weight is allowed to fall on the top of the drill rods until the sampler has penetrated about six-in into the soil, whereupon the penetration test is started and the foreman records the number of blows to produce the next foot of penetration.'

Reference to Hvorslev (1949) indicates that some tests were carried out without a 6-in seating drive, simply starting the test 'from the depth to which the sampler sinks under the weight of the drill rod'. According to some sources (e.g. Schnabel, 1966) the standard test developed by the Raymond Pile Company required that 'the sample spoon be seated on the bottom of the hole with a few light taps prior to driving and recording the number of blows for 1ft. penetration'. In 1954, Parsons introduced the procedure of recording the number of blows for each of three 6-in increments of penetration, adding the last two to give the SPT N value (Fletcher, 1965). By 1958, the ASTM tentative standard described the seating drive as follows:

> 'with the sampler resting on the bottom of the hole and the water level in the boring at groundwater level or above, drive the sampler through undisturbed soil 6 in. with a few light taps. Then drive it 12 in. or to refusal by dropping the 140 lb. hammer 30 in. Record the number of blows required to effect the 2nd and 3rd 6 in. of penetration.'

Since 1948 the emphasis in this sampling/testing technique has changed. As described above, in the early days the sampling process was of primary importance, with the penetration resistance providing additional, relatively inexpensive, information. With the development of our understanding of the factors controlling sampling disturbance, however, it has been widely appreciated that any sample obtained from the 2-in sampling tube used with the SPT is highly disturbed, and in some soil types may not even be representative. In most countries, therefore, the primary objective of the SPT procedure is now to obtain the penetration resistance of the soil.

In the UK, small-diameter boreholes have not been widely used for site investigation purposes, because as Harding (1949) noted, 'in British gravel-laden deposits, nothing less than 6-in diameter is worthwhile. This permits the average type of stone to be brought up by shell without pounding with a chisel and also allows of 4-in diameter sampling'. Open-drive sampling evolved in the United Kingdom in the 1930s and 1940s based upon much larger diameter sampling tubes than in the USA (Le Grand *et al.*, 1934) so that by the time the SPT procedure was introduced into the UK the U4 (now known as the BS general purpose sampler or U100) was well established; the SPT has thus always been regarded in Britain as an *ad hoc* penetration test rather than an exercise in sampling. Relatively soon after its introduction into the UK the test was modified by the use in gravels and stony soils of a 60-degree conical point instead of the shoe of the split spoon (Palmer and Stuart, 1957).

The SPT has been used in almost every part of the world. The test was first standardised in the USA in 1958 (ASTM Designation D1586-58T). It was introduced into BS1377:1967 as the 'Determination of the Penetration Resistance using the Split-Barrel Sampler'. It is currently covered by ASTM D1586-84 (Reapproved 1992) and BS1377:Part 19, 1990, and by many other standards (see Section 4.1). The *Report of the Sub-committee on the Penetration Test for use in Europe* (ISSMFE, 1977) provides the basis of a standard for Europe. More recently (1988) an International Reference Test Procedure (IRTP) was published at the 1st International Symposium on Penetration Testing under the auspices of the International Society for Soil Mechanics and Foundation Engineering (ISSMFE, 1988).

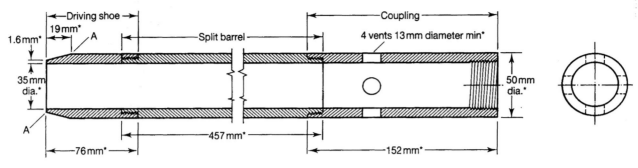

Corners at A might be slightly rounded.

This design has been found satisfactory, but alternative designs may be employed provided that the essential requirements are fulfilled. (Essential dimensions are indicated by an asterisk.)

(a)

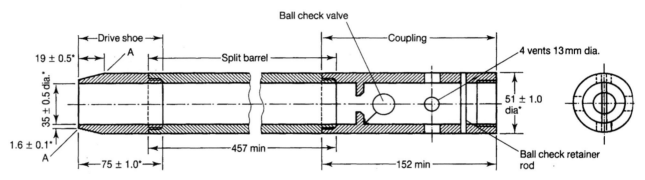

Corners at A may be slightly rounded.

This design has been found satisfactory, but alternative designs may be used provided that the essential requirements are fulfilled.

Figure 4 *BS 1377 split-barrel samplers*
(a) BS 1377: 1975 (b) BS 1377: 1990

However, as noted by Ireland *et al.* (1970), one of the main problems of the Standard Penetration Test remains that neither the test equipment nor the associated drilling techniques have been fully standardised on an international basis. Differences in equipment and technique exist partly because of the level of capital investment current in different countries, but more importantly because of the adaptation of drilling techniques to local ground conditions. These differences lead to variations in the test result, a situation which is clearly undesirable because of the empirical usage of the SPT N value in geotechnical design.

A synopsis of the SPT's evolution is as follows:

1. The Standard Penetration Test was introduced in 1947, and is now in widespread use because of its low cost, simplicity and versatility.

2. The SPT procedure initially arose from a desire to obtain cheap additional information during small-diameter sampler driving.

3. The test is not yet fully standardised, either nationally or internationally.

4. Differences in boring, equipment and test procedure are likely to influence the SPT N value.

5. The SPT provides a simple, universally applicable, testing method.

6. No sophisticated boring or testing rig is required.

7. The test has sufficient versatility that it can provide information on hard-to-sample soil and weak rocks.

3 UK Practice

3.1 THE BRITISH STANDARD METHOD OF TEST

The current British Standard is BS 1377:1990 Part 9, Section 3.3 (*Determination of the penetration resistance using the split-barrel sampler (the Standard Penetration Test, SPT)*). Figure 4 shows the split-barrel sampler design given in BS 1377. Figure 5 shows, in abbreviated form, the main provisions of the British Standard. It remains to be seen how the Standard will affect British practice; indeed the provisions of BS 1377:1975 for the SPT had not been fully implemented in its fifteen years of currency.

The principal changes made by the new standard are:

- a requirement, following a reduction in casing diameter, to drill for at least 1m before carrying out an SPT

- a return to a 63.5-kg (140lb) drive weight

- the specification of an anvil mass between 15kg and 20kg

- a specified maximum overall weight for the drive assembly (or trip hammer) of 115kg

- the introduction of a maximum rod weight of 10kg/m

- the specification of a maximum permitted curvature for bent rods, in the form of a relative deflection of 1/1000

- the introduction of a ball check-valve in the head of the split barrel (or split-spoon) sampler assembly (Figure 3)

- the seating drive should follow any initial penetration under the dead weight of the split spoon, the rods and the hammer

- redefinition of the procedure for driving the split-spoon, including optional reporting of the blow counts for each 75mm of penetration

- maximum blowcounts for the seating drive and test drive, as:

	Maximum blows to be applied:	
	in seating drive	in test drive
Soil	25	50
'Soft rock'	25	100

- where the penetration under the dead weight of the rods and hammer exceeds 450mm, the N value is recorded as zero.

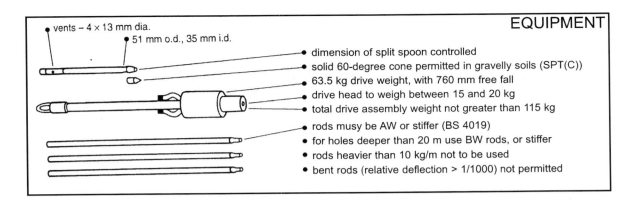

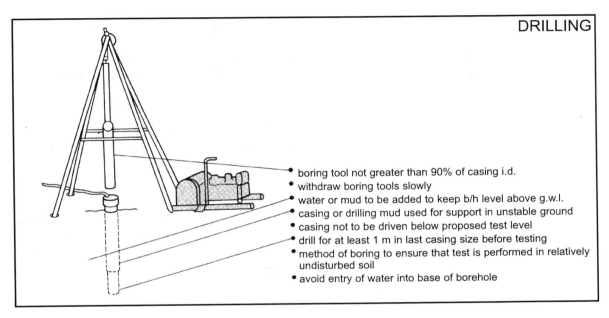

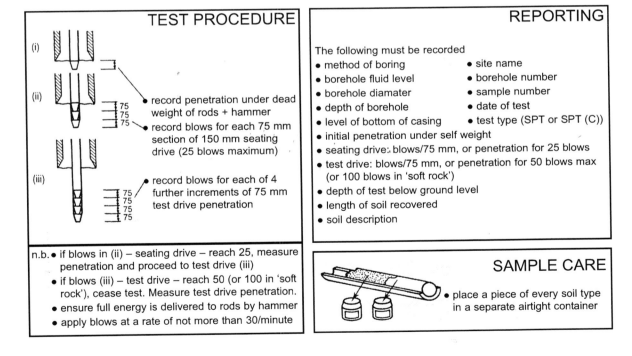

Figure 5   *British Standard requirements for the SPT*

3.2 UK BORING TECHNIQUES

In the UK, the borehole is usually advanced using a light percussion boring rig, by claycutter or cross-cutter in clays and by shell (sometimes termed a bailer) in sands or gravels (Figure 6).

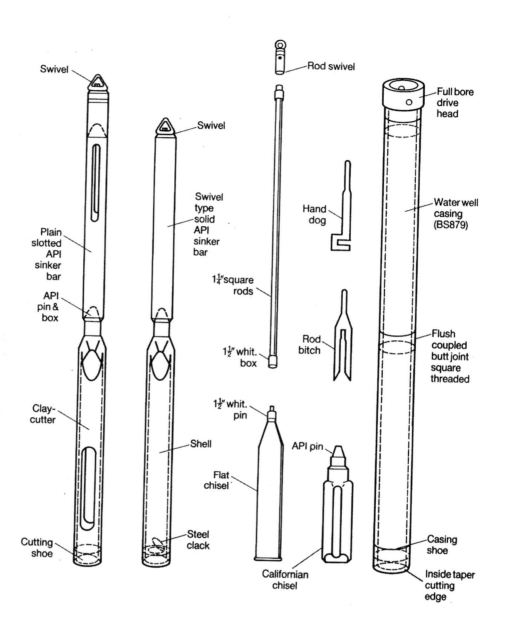

Figure 6   *Light percussion boring equipment (after Clayton et al., 1983)*

The *minimum* disturbance that is produced, regardless of the technique used to make a borehole, comes from the reduction of vertical total stress at the bottom of the borehole. If the borehole is dry, then the vertical total stress is reduced to zero. If the borehole is maintained full of water or mud, at all times, then the magnitude of vertical total stress relief can be halved. The depth of soil significantly affected, below the level of the bottom of the borehole, is approximately equal to 1.0 – 1.5 times the borehole diameter.

The claycutter is weighted with a steel sinker bar which weighs about 60 kg, and this assembly (up to 150kg) is dropped from a few metres above the bottom of the hole. It embeds itself in the soil, and the rig operator then uses the rig winch slowly to pull it out, and lift it to the surface. Clay which has become lodged inside the tool is removed by a second man, who levers and drives it out by inserting a steel bar into the open slot in the side of the claycutter. Provided that the claycutter is not dropped from too great a height, and is not used in soft or very soft soil, the amount of disturbed soil below the base of the borehole can be expected to be small enough not to influence the result of an SPT.

In granular soil a shell is used, and at least 2m of water must be present in the bottom of a borehole for it to function. The shell is lowered to the bottom of the hole, and rapidly raised and dropped, moving about 150mm up and down every second or so. Lifting the shell upwards causes water to be drawn into the base of the borehole, forcing some of the granular soil into suspension. As the shell is dropped to the bottom of the hole, the non-return flap-valve at its base (sometimes termed a clack, see Figure 6) opens and a mixture of soil and water passes into the body of the shell. As the shell is subsequently raised, the clack closes as water tries to flow out. The coarser soil particles then settle on to the top of the clack.

By repeatedly pumping the shell up and down at the bottom of the borehole, granular soil is gradually collected in the barrel of the shell so that it can be removed from the hole. The casing which supports the sides of the borehole can be allowed to sink under its own weight if it is free, but more often it is driven gently downwards as the borehole is advanced, or surged (i.e. raised and dropped). It may also be lowered more gently by rotating it by hand, with chain tongs. The fastest method of drilling (and therefore the most profitable) often uses a tight-fitting shell. This causes the groundwater to flow more rapidly into the base of the hole and, because hydraulic gradients in the soil are high, induces the soil to pipe (or 'blow') up the casing. Even when water is continually added to the hole, the removal of the shell quickly from the borehole causes a sudden drop in the borehole water level. Begemann and de Leeuw (1979) calculate that a drop in water level of only 0.34m is sufficient to cause piping in granular soil.

Piping causes soil to flow up inside the casing, and in the process loosens the soil below and around the borehole. SPTs carried out in soil which has piped will give lower penetration resistances than would be obtained from undisturbed soil. It is thought that this process can disturb the soil to a depth of about three borehole diameters below the base of the casing (Reidel, 1929). Thus when using a shell with a 6-in. (152 mm) or 8-in. (204 mm) diameter casing, the soil can be loosened to a depth at least equal to the test depth of the SPT. In practice, deep boreholes start with larger diameter casing (as great as 15-in. or 375 mm diameter), and the disturbed zone can then be even deeper.

A further problem that can arise, from a combination of poor drilling technique and a large borehole diameter, occurs as a result of poor control of the level of the bottom of the casing relative to the current level of the bottom of the borehole. In normal good practice the casing is never allowed to go below the base of the borehole. Driving casing ahead of the borehole forms a plug, and the plugged casing then tends to push the soil aside rather than allow it free entry inside the hole. Compaction and remoulding can then occur for up to three times the diameter of the casing, below its final position (Hvorslev, 1949). Once again, this problem is particularly pronounced in granular soil.

Rotary drilling is sometimes used to form holes in which SPTs will be carried out. The technique is normally used in the UK only in rocks, but some engineers prefer the quality of sample that can be obtained in weak rocks and heavily over-consolidated soils when high quality rotary coring is used. Rotary holes may also be advanced (for example, through soils overlying rock) without taking core (termed open-holing). In rotary drilling the soil or rock is ground into very small particles (cuttings) before being lifted to the ground surface by a flush fluid. This fluid may be air-based, but it is normally 'mud' (a mixture of polymer and water, or of bentonite and water) or simply clean water. Because the cuttings must be brought to the top of the hole, to avoid blocking off, it is necessary to maintain an upward flow of flush fluid at all times. The hole is therefore guaranteed to be full of fluid, and if bentonite mud is used the fluid will be denser than water. In addition, many drilling fluids avoid the need for casing. These factors mean that inflow of water into the hole is unlikely, except when pulling the tools from the hole, and that the concentration of upflow of groundwater into the base of the hole can be avoided. The drilling bits in use conventionally use downward facing flush ports, which may cause disturbance to the bottom of the hole in loose uncemented granular soils.

The provisions made by the various standards and specifications to reduce disturbance to the SPT test section are:

- maximum allowed weight for claycutter and sinker bar
- shell or claycutter diameter not greater than 90% of internal diameter of casing
- borehole water or mud level to be maintained above groundwater level at all times during boring
- casing not to be driven below the top of the SPT test section
- bottom discharge bits are not permitted if washboring is used.

Current UK practice (as distinct from the British Standard procedure) ignores some of the most important of these provisions. Undersize shells are not always used, even in the loose fine sands that are so prone to boil. Water is normally only added to boreholes 'to assist boring', and the driller will generally only determine the groundwater level once each shift, by measuring its level in the hole at the start of the day's operations. He has little knowledge of the equilibrium ground water level during boring. On the positive side, casing is rarely driven below the bottom of the hole.

The process of boring or drilling a hole requires great skill, and the soil disturbance created depends largely upon the technique of the rig operator, but also upon the soil type, groundwater and borehole fluid levels both during boring down to the test section and during the test itself, and on the equipment used. The importance of boring technique becomes obvious when it is possible to compare the relative volumes of granular soil removed from similar boreholes by different drillers. The effects of drilling on the SPT N value are therefore both extremely variable and unpredictable. But the discussion above suggests that while the effects can be severe in silts, sands, gravels and very soft or soft clays, they will be limited in other cohesive soils.

3.3 UK STANDARD PENETRATION TEST EQUIPMENT

Usual practice in the UK is to use a split-spoon sampler with no provision for liners, no inside clearance, and no ball-check valve. Samplers usually comply with all the requirements of BS 1377:1990 (see Figure 3), except that through over-use, or as a result of hitting stones, the cutting shoe becomes blunted and burred, and a ball check-valve has not so far been in use. Split spoons with liners can be obtained, but are rarely used. A 60° solid cone is often screwed on the end of the split spoon in place of the cutting shoe when the test is made in gravel or gravelly sand, or in stoney materials such as glacial till or chalk. The test is then designated SPT(C) or SPT(cone) on the records of the borehole. The use of a solid 60° cone was permitted, but not mandatory, in BS 1377:1975, and this continues to be the case in the current standard. Plastic core catchers are available, but are not often used.

The sampler is connected to the driving mechanism at the top of the borehole by rods which, according to BS 1377, should have a stiffness equal to or greater than a US Diamond Core Drilling Manufacturer's Association (DCDMA) AW rod (BS 4019: Part 1), which has an outside diameter of 44 mm. In many cases, British light percussion boring rigs are still equipped with the traditional solid square boring rods, which are nominally 1.25 × 1.25-in in cross-section. Such rods were very widely used until the mid-1970s, both for chiselling and for SPTs carried out at depths to about 40 m. Below this depth solid 1.5-in × 1.5-in rods were used, to avoid buckling under the self-weight of the rod string.

Figure 7   *Automatic trip hammer in use*

At present, some rigs use hollow, round, rotary-drilling rods, many of which comply with the standard. The lightest rods known to be in use are A rods, whilst the use of rods as heavy as NW has been reported (Mallard, 1983). According to the British Standard, AW rods should be supported by steadies when testing at depths greater than 20m, or more rigid BW rods should be used; in practice steadies or stiffer rods are rarely used, except for very deep holes.

The automatic trip hammer, or 'monkey', is used as standard in the UK (see Figure 7). This hammer was first devised in 1963 by the Piling and Construction Company (later Pilcon Engineering Ltd) and rapidly gained acceptance because of its rugged construction and simplicity in use (Stubbings, 1966). Two versions are currently in use; that by Pilcon, and a second but similar type produced by Duke and Ockenden Ltd (Dando) (Figure 8).

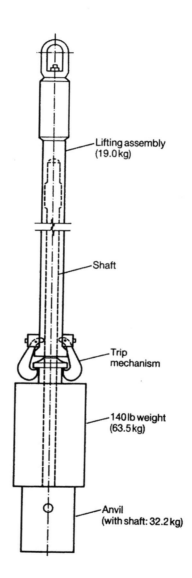

Figure 8   *British automatic trip hammer*

The tool consists of three components:

1. An inner stem connected at its base to an anvil ('drive-head' or 'striker plate') which screws directly to the drill rods. Current British anvils are large, weighing up to 20kg.

2. A 140-lb cylindrical weight which has a small central hole allowing it to slide freely over the inner stem.

3. An outer tube assembly with pawls at its base. These pawls pick up the 140-lb weight and release it when the weight has been lifted 30-in above the upper surface of the striker plate.

This type of hammer is particularly desirable because it delivers the blow in a consistent manner, unlike other techniques in common use in other countries. It should be noted, however, that British automatic trip hammers continued to use a 140-lb (63.5 kg) weight, rather than the 65 kg weight which was required by BS 1377 from 1975 until the introduction of the current standard.

3.4 TEST PROCEDURES IN CURRENT USE

BS 1377:1990 requires that the borehole shall be cleaned using equipment which ensures that the material in the test section is 'relatively undisturbed'. As noted in Section 3.2, common practice in granular soils is to use a full-diameter shell or bailer. The maximum depth of the shell is rarely checked, but it is common for the driller to lower a weighted tape to check the level of the bottom of the hole. The depth of the bottom of the casing is normally estimated from the number of casing lengths in use and the stick-up at ground level. While the level of the bottom of the hole can be calculated, the thickness of disturbed soil or material fallen from the sides of the hole will not be known.

The split-barrel sampler is then lowered on rods to the bottom of the hole and the trip hammer assembly connected. The SPT tool penetrates under the dead-weight of the rods and trip hammer, before being driven the 150mm of the seating drive. The blows for the seating drive are recorded for two 75mm increments of penetration, although this is optional in the current British Standard. It is commonly assumed that the 150mm seating drive penetrates completely the zone of soil disturbed by the action of boring, but in some cases (for example, in granular soil) this may not be so. The sampler is then driven a further 300mm, sometimes recording the blows for each of four 75mm increments of penetration, sometimes for two 150mm increments, and occasionally even for a single 300mm drive. The rate of application of blows is not standardised, but may typically be about 15-20 blows per minute. The total blows for the last 300mm of penetration is the penetration resistance, N. When values are obtained using the standard cutting shoe, the result is recorded either as *SPT* or as *SPT*(S) on the driller's daily report. When the solid 60° cone is in use, the test is recorded as *SPT* (*cone*) or *SPT*(C).

Test procedures are varied for very hard soils, and sometimes for very soft or loose soils. In the case of hard soils, BS 1377 recommends a maximum of 25 blows for the seating drive, and 50 blows for the test drive in soils, or 100 blows for a test drive in 'soft rock', but no further detail is required other than the total penetration for these blow counts. A good practice, used by some contractors, is to record the penetration for increments of 20 blows. The previous British Standard did not give guidance on what should be done if more than 50 blows were used during the seating drive, and as a result many different procedures are adopted. The maximum number of blows to be given in an entire test (i.e. including the seating drive) varies from one investigation to another. Some drillers use a maximum of 50 blows, while other use 80 or even 100 blows, and there are often inconsistencies between drillers working for the same company, or between different sites investigated by the same specialist contractor.

In very loose or soft soils the weight of the rods plus the trip hammer alone may cause the SPT tool to drop through part or all of the test section. Some operators record this as a zero penetration resistance. In other cases the tool is allowed to stabilise under its self weight before the seating drive (and therefore the blow count) is started. In very loose soils, where 15 blows or less achieve the total penetration of 450mm, some engineers require that further penetration should be made until either the total blowcount exceeds 80 at the end of a penetration increment, or the total penetration exceeds 750mm. This procedure may be helpful in demonstrating that boring disturbance has occurred, but its results should be clearly marked to distinguish it from the standard procedure.

BS 1377:1975 (test 19 note 4) stated that 'it is usual practice to modify the measured value of penetration resistance on the basis of the empirical rule described by Terzaghi, K. and Peck, R.B.', i.e. $N' = 15 + 0.5(N-15)$ when the N-value is greater than 15 in saturated fine or silty sands. This has not been normal practice, and the new standard has abandoned this recommendation.

3.5　SYNOPSIS

1. The recommendations of BS 1377:1975 and BS 1377:1990 Part 9 have not been followed in practice. In some cases (for example, minimising borehole disturbance) it is doubtful if they could be.

2. The effect of varying equipment and practice is to produce variations in SPT N value.

3. Good supervision of SPT operations, and training of rig operators, is required in order to achieve consistent results.

4. There is a need to standardise procedures for tests carried out in weak rock, because this has not been completely achieved in BS 1377.

5. In granular soils, additional information can be obtained when the penetration resistance is low, by driving the split spoon through further 75mm increments.

6. Although drilling disturbance may have a considerable effect on SPT results obtained in granular soils, for most cohesive soils it will not be of major significance.

4 Practice outside the UK

4.1 NATIONAL STANDARDS

The first tentative standard for the SPT was published in April 1958 by the US American Society for Testing and Materials (ASTM); this became a full standard in 1967 (ASTM D1586 – 67) and was re-approved in 1974. The current editions, D1586 – 84, was re-approved in 1992 Figure 9 shows the ASTM standard split-spoon sampler.

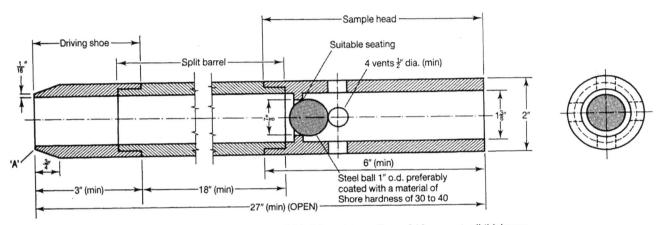

BS1377-9 1990
OR 2008.

Figure 9   *ASTM split-barrel sampler (ASTM D1586-84, re-approved 1992)*

The ASTM standard is broadly similar to the current British Standard, except that:

- A rods (1-⅝in. (41.2mm) o.d., 1-⅛in. (28.5 mm) i.d.), which are slightly lighter than AW rods, or stiffer rods to N size are permitted for holes to 30m depth

- the borehole diameter must be between 2-¼ in (57.2mm) and 6.5in (162mm)

- the use of a cathead to pull a rope attached to the hammer

- the use of 1.5mm thick liners is permitted, provided that the internal diameter of the sampler remains 1-⅜in (34.9mm)

- core catchers may be fitted in the driving shoe to prevent sample loss.

ASTM D1586-67 describes the stages of the test itself as:

1. Rest the split spoon on the bottom of the borehole.

2. Drive the split spoon with the 140-lb. hammer falling 30in and count the number of blows applied in each increment until either a total of 50 blows in any one increment, or a total of 100 blows is reached, or there is no observable penetration, or the full penetration of 18in is achieved.

3. Repeat the procedure every 5 ft (1.5m) or closer in homogeneous strata, and at every stratum change.

4. Record the number of blows for each 6 in (0.15m) penetration. If a full drive is achieved, add the blows for the last 0.30m of penetration, to give the penetration resistance N. If not, the number of blows for each complete or part increment shall be recorded on the boring log. Part increments of penetration shall be reported to the nearest 1in.

Table 1   *National standards for the Standard Penetration Test*

Country	Own standard	Use of other standard
Argentina		ASTM D1586
Australia	SAA Test 16A (1971)	–
Brazil	NBR 6484 - 1980	–
Canada	CSA A 119.1 - 1966	–
Czechoslovakia	CSN 73 18 21	
Egypt		ASTM D1586 and 1377
Greece	–	Earth Manual (USBR, 1963)
Hong Kong	–	BS 1377:1975
India	IS:2131 - 1963	–
Israel	–	ASTM D1586
Iraq	–	ASTM D1586
Italy	–	ASTM D1586
Japan	JIS 1219, 1976	–
Mexico	–	ASTM D1586
Nigeria		BS 1377:1975
Norway	–	Terzaghi & Peck (1948)
Poland	*	
Portugal	*	
Saudi Arabia		ASTM D1586 and 1377
South Africa	–	ASTM D1586
Spain	–	Terzaghi & Peck (1948)
Switzerland		ASTM D1586
Turkey	TS 1900 - 1975	
United Kingdom	BS 1377:1975	–
United States	ASTM D1586-67	–
Venezuela		ASTM D1586

* standard thought to exist, but designation unknown.

Table 1 lists countries with their own or adopted standard specifications for the SPT. The main difference between these other standards and the current British one are:

- variations in hammer weight and drop as a result of metrication of the original procedure
- limits on borehole diameter, typically between 2-¼in (57mm) and 6in (152mm)

- the methods of boring which are permitted, and the details of borehole fluids and their required levels
- the use of ball check-valves, and the differences in vent sizes in the sampler head
- the permissibility of a 60-degree solid cone. Only five countries (UK, South Africa, Australia, Portugal and Spain) are thought to use the cone, and of these Australia, Portugal, and the UK have their own national standard
- the option of using liners and core retainers
- the maximum number of blows to be used before stopping the test, which varies between 50 and 100 blows
- the test records which are required. As an extreme, the Japanese require that the penetration for every blow should be recorded.

Table 2 shows in detail the provisions of the standards used in 11 countries, as found by the ISSMFE SPT Working Party in 1988.

Where a country does not have its own national standard it is usual for consulting engineers to adopt another's standard, often that of the ASTM. In practice, however, the tests have to be carried out using whatever equipment is locally available.

4.2 THE ISSMFE EUROPEAN AND INTERNATIONAL REFERENCE TEST PROCEDURES

From 1957 to 1965, sub-committees of the International Society for Soil Mechanics and Foundation Engineering (ISSMFE) considered the methods of static and dynamic penetration testing then available, with a view to standardisation. Following a wide divergence of views at the Montreal Conference in 1965 the ISSMFE sub-committee was dissolved, but the European national societies, wishing to continue this work, set up a European sub-committee. They produced, at Tokyo in 1977, the *Report of the Subcommittee on the Penetration Test for use in Europe*. This document gave recommended 'standard tests' selected from the wide variety of penetration tests in use at the time in the region, and presented Standards for them in English and French. One of the standards given in the Appendix to that document was for the SPT, and was based upon ASTM D1586-67 (re-approved 1974) and BS 1377:1975.

In 1982 the decision was made at the Second European Symposium on Penetration Testing (ESOPT II) to form a Technical Committee of the ISSMFE to report on penetration testing. The idea was conceived of an International Reference Test Procedure (IRTP) as opposed to an international standard. An IRTP aims to give the simplest complete description of all the features influencing the test result, while recognising as far as possible the need to allow local variations to cope with special ground conditions. Appendix 1 gives the ISSMFE Reference Test Procedure for the SPT as published at the 1st International Symposium on Penetration Testing (ISOPT I) in 1988.

Table 2   *SPT standards in different countries*

Legend: ● = as draft reference procedure   ○ = not specified   * = information in note

Main items of specification	Australia	Canada	Czechoslovakia	Greece As earth manual 2nd ed. 1974	Japan	Mexico As ASTM 1586 dated 1967	Portugal	United Kingdom (BS 1377 : 1975)	USA (ASTM 1586–67) reapproved 1974	Poland	India
Scope											
Hammer wt (63.5) kg	●	●	●	●	●	64	64	65	●	65	65
Drop (760) mm	●	●	750	●	750	750	750	●	○	750	750
Boring restrictions											
Clean hole capability	●	●	○	●	●		●	●	●		●
Side discharge bit only	●	●	○	●	○		○	●	●		○
Bailer dia. restricted	60%	○	○	○	do not disturb test zone		○	90%	○		● prefer casing turned
Casing/mud in weak soils	●	●	○	○	●	casing	casing	●	●		
Hole dia. restriction mm	○	57-152	separate standard for site exploration	57-152	65-150		○	○	57-152		55-150
Sampler assembly											
length mm	● ● min 533	● ● min 533	● ● 535	● ● min 483	● ● 535			50 ● ● 533	● ● 533		50.8 ● ● 675
Ext/int. dia. shoe & sampler	● ●	● ●	● ●		● ●	○	no details	● ●	● ●		● ●
Drive shoe separate	387 ●	● ●	● ●		not specified	●		390 ●	● ●		● ●
Vent area mm² / check valve											
Liner permitted	yes	yes*	no		no	yes		no	yes		yes
Split or tube	either	split	split		split	tube		split	split		split
Hardened steel shoe	●	●	○		○	●		●	●		?
Solid cone	yes	no	no		no	○		yes optional	no		no
Other	optional thin core retainers	optional – core retainers	–		–	–		–	optional core retainers		–
Drive rods											
depth m					40.5 o.d. (just under A)	> 15 m	no details		stiffer A over 15 m	42/51 dia.mm	A min
size, range dia. mm	AW/N	AW/N	○	AW/B	AW/NW			AW/min	no		
steadies, start/spacing (m)	15¼ 6 (or stiffer rods)	no	no	○	no			15 3 (or stiffer rods)		10	–
straight/tightly coupled	○ ○	● ●		○ ●	○			○	○		
Drive weight											
Guide/freefall	● ●	● ●	● ●	● ●	● ●	cathead	no details	● ●	● ●		● ●
Trip mechanism	● * doomed	○ no mention	○ advised	● advised	● permitted	○		● mandatory	○	18 kg	○
Anvil specification		○	○	jar coupling	● 75 o.d.	○		● not defined	○		○
Preparing borehole											
clean hole	●	●	○	●	●	do not disturb test zone		●	●		●
positive head	●	sands & silts	○	●	○			●	○		○
Bailer raise slowly	●	●	○	●	○			○	●		●
casing at/above test level	●	●	○	●	●			●	●		●
Executing test											
initial gravity penetration (I.P.)	○	○	○	○	○	○	○	○	○		○
high IP/omit SD	○	○	○	○	○	○	○	○	○		○
excessive IP/omit N	○	○	○	○	○	○	○	○	○		○
SD max. blows (50)	●	●	○	●	●	●	●	60	●		○
SD procedure	●	●	light blows only	●	●			●	●		
max blows	60 incl. SD	100 incl. SD	○	50 incl. SD	50 during N test	50 per 100 mm		100 incl. SD	50 during N test		20 per 250mm
Blow rate	○	○	○	○	○				100 incl. SD or 50 for 50mm or less		○
Soil sample									○		
sample retention	●	●	●	●	●	●	●	●	●		●
sample labelling	●	●	●	●	●	●	●	●	●		●
Reporting											
boring method	●	●	●	○	○		○	●	●		●
rod size	○	○	○	○	○		○	○	○		○
hammer type	○	○	○	○	○		○	○	○		○
initial gravity penetration (L)	●	●	●	○	○		○	○	○		○
SD short											
SD / TD blows/L (mm)	blows every 152	– every 152	no record every 100 advised	no record 300*	no record Penet./blow but if penet. <20, report blows/100		blows & penet. 150 300	150 every 75	every 152		every 150
TD procedure (N)	●		●		●		●				●
short test drive	–	advance for final 20 blows	–	penet./50 blows	–		depth base casing	blows & penet.	blows & penet.		N=last 300mm
casing data	casing size	casing used	○		○			casing size	casing used		

The International Reference Test Procedure is basically the same as the British Standard, with the following exceptions:

Boring: The diameter of the borehole must be between 63.5 and 150 mm.

Equipment:

1. The use of a solid 60° cone was not approved.

2. Rods heavier than 10.03kg/m are not permitted.

3. The overall weight of the hammer assembly must not exceed 115 kg (250 lb).

4. Suitable weights and section modulus values are given for the drive rods as follows:

rod diameter (mm)	section modulus ($m^3 \times 10^{-6}$)	rod weight (kg/m)
40.5	4.28	4.33
50	8.59	7.23
60	12.95	10.03

Note that the AW, BW and square rods permitted by the British Standard fall within this range, having the following properties:

size	rod diameter (mm)	section modulus ($m^3 \times 10^{-6}$)	rod weight (kg/m)
AW*	43.6	5.10	4.57
BW*	54.0	10.12	5.80
Sq.	31.8 (1 ¼in)	5.33	7.89
Sq.	38.1 (1 ½in)	9.22	11.33

* Diamond Core Drill Manufacturers' Association (USA)

Test procedure: The maximum rate of testing should be 30 blows per minute.

4.3 INTERNATIONAL PRACTICE AND EQUIPMENT

There are significant differences between the drilling techniques, SPT equipment and test procedures used in different countries. These differences have arisen because of adaptation to local ground conditions, because of the relative difficulty of access to borehole sites, and because of the degree of sophistication of drilling and boring plant. They can have a major influence on the result of a Standard Penetration Test.

The main variables are:

- method of drilling
- borehole fluid
- borehole diameter
- hammer mechanism

- rod stiffness
- split-spoon geometry
- method of testing.

The methods of drilling most widely used around the world are wash-boring, augering, rotary drilling, and light percussion drilling.

Wash-boring is widely used in South America and South Africa, sometimes used in North America, and only rarely used in the United Kingdom. In this method progress is normally made by dislodging soil from the base of the borehole by the chopping and washing action of a cutting bit which is raised and dropped at the base of the hole as flush fluid is pumped out of it. If the bit has flush ports which face downwards then disturbance ahead of the borehole is to be expected, so some standards, including BS 1377: 1975 contained clauses to prevent this. The current British Standard permits wash-boring provided that a side discharge bit is used. The method has the advantage that the hole is maintained full of drilling fluid (normally water) at all times, so that blowing is unlikely. This technique is normally used to make small diameter boreholes (generally less than about 100mm diameter) and so (in contrast with the large diameter holes used in the UK) the depth below the bottom of the hole influenced by stress relief will be small. However, it is difficult to use in British soil conditions because coarse granular material tends to be left at the base of the wash boring. In such conditions there is a need to chisel, and a danger that non-representative samples may be taken. *In-situ* tests such as the SPT, therefore, might be carried out in or through unrepresentative borehole debris.

Augering is often used in North America and the Middle East, and is sometimes used in the United Kingdom. Lorry-mounted continuous hollow-stem flight augers are gaining popularity worldwide, because of the rapid rates of boring and drilling that can be achieved. This type of rig has failed to gain widespread acceptance in the UK because it is expensive, and requires much better access than is normally available to the site of British boreholes. Fears have also been expressed about the degrees of drilling disturbance. Auger rigs are often capable of exerting a hydraulic downthrust on their drilling tools of several tonnes. This thrust can be used to increase the rate of drilling, but in so doing, or if the auger tends to drill itself into the ground, the soil ahead of the auger will be remoulded and displaced (if cohesive) or compacted to a higher density (if granular).

In a solid-stem auger system the augers must be pulled out of the borehole before the SPT tool can be lowered to the test depth. The action of raising the augers creates suction at the base of the hole, which may encourage boiling. If the unsupported hole collapses then there is the possibility of the test being made through the caved-in and remoulded soil. When hollow-stem augers are used, the plug at the base of the hollow stem has to be removed to allow access for the SPT split spoon to the base of the hole. Any suction created by removing the plug, or drop in water level as the rods a removed, may induce boiling and loosening of the soil in the test section. The penetration resistance will be reduced if the test is carried out below the augers, but will be increased if the driller fails to notice that the hollow stem has partially filled with soil and makes the test within the auger.

Rotary drilling techniques are generally used with the SPT in Japan and Hong Kong. In Japan a bottom-discharge rotary action bit is used, which is forced into the ground by the action of a hand lever (Muromachi *et al.*, 1974). The stability of the borehole is maintained with drilling fluid, usually bentonite mud. In other countries, and in relatively hard ground conditions, rotary drilling techniques with water flush and casing are used. Progress is made with bottom discharge bits, rock roller bits, or drag bits, generally in small diameter holes. When loosening of the test section is avoided, by specifying upward deflecting drill bits, this method produces good results (Seed *et al.*, 1985).

The borehole fluid most widely used is water. In the UK light percussion boreholes are rarely completely full of water but, as noted in Section 3.3, this is largely because of the difficulties of the method of drilling rather than ignorance of the consequences. Water is the normal flush fluid used for wash-boring, e.g. in Brazil. Some modern drilling techniques, such as hollow-stem augering use no borehole fluid at all. The potential for borehole stress relief and boiling is then large, particularly when the central plug is removed from the augers prior to testing. Schmertmann (1966) noted the advantages of drilling mud as a borehole fluid for SPT test holes, and this seems to be the current practice in Japan (Kovacs and Salomone, 1984) where mud is used with a bottom-discharge bit in a rotary wash-drilling method.

Schmertmann (1978) has commented that 'the use of rotary drilling methods with the hole continuously filled with drilling mud to the surface offers the only present way to assure that the effective stress conditions in the sampling zone immediately below the borehole remain as little disturbed as possible by the borehole'. Drilling mud is not often used and, indeed, the technique is little understood, by ground investigation drillers in the UK. In wash-boring or rotary drilling, where the mud can be recirculated, very little water need be brought to the borehole, which makes drilling easier and more economical. An advantage in standard penetration testing is that borehole support is given without the use of casing, and so blowing and loosening of the base of the hole is less likely. Unfortunately, there seems no practicable way of recirculating the drilling fluid in a light percussion borehole.

Borehole diameter varies from as little as 60mm to as much as 375mm. The original American boreholes, described by Fletcher (1965), were wash-bored using 2.5-in (63.5mm) or 4-in (101.6mm) internal diameter casing. Typical borehole diameters in current use are 100-150mm (Israel), 65-110mm (Japan), 55-150mm (India), and 60-76mm (South Africa). In contrast, British practice typically uses 6-in (152mm) and 8-in (204mm) casings and may use sizes up to 15-in (375mm).

Borehole diameter, in itself, is probably not of serious consequence provided stress relief at the bottom of the hole can be kept to a reasonable level, and boundary pore pressures can be maintained. Unfortunately, neither factor can be controlled during light percussion or auger boring. The borehole diameter then becomes important in controlling the amount and depth of disturbance. Light percussion drilling, as practised in the UK, then looks poor in comparison to the techniques used in other countries.

In most countries the specified drop and mass of the hammer have not been changed by metrication, and remain 2ft 6in and 140lb respectively. Hammer mechanisms can be divided into the following groups:

- automatic trip hammers
- hand-controlled trip hammers
- slip-rope hammers, and
- hand-lifted hammers.

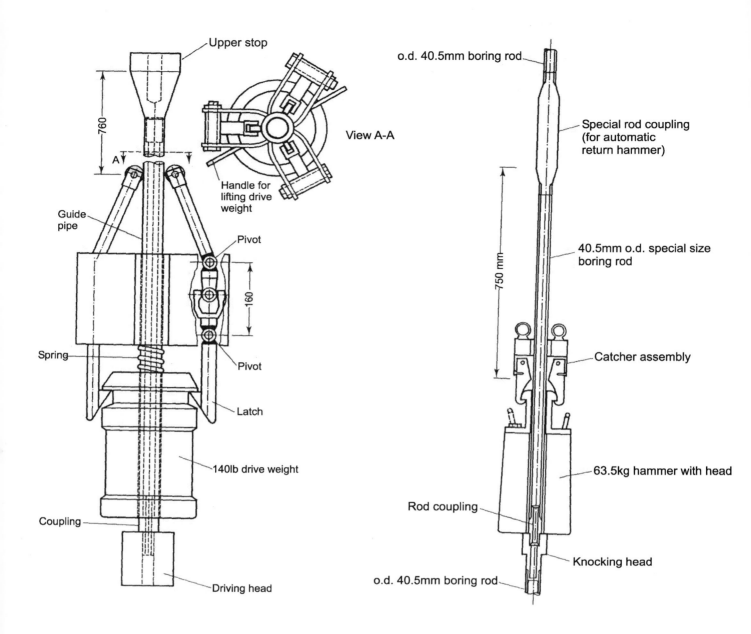

Figure 10 *Automatic trip hammers (a) Israeli (Komornik, 1974), (b) Japanese (Japanese Society of Soil Mechanics and Foundation Engineering, 1983)*

Automatic trip hammers, which are standard in the UK, are also used in Israel, and Australia. Figures 10a and b show the trip hammer designs used in Israel and Japan, which can be compared with those in use in the UK (Figure 8). Automatic trip hammers are obviously to be preferred in Standard Penetration Testing since they ensure that the weight falls freely and consistently through the correct distance. They do not, however, ensure that the theoretical free-fall energy is delivered to the rods. Differences in anvil weight are thought to have a major effect on hammer efficiency (for example see Skempton, 1986), and it is believed that if the hammer is not vertical during testing, then less energy will be delivered as a result of frictional losses.

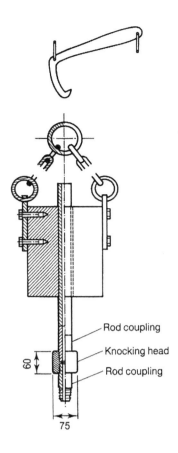

Figure 11 *Japanese hand-controlled SPT trip hammer components (after Muromachi et al., 1974)*

An example of a hand-controlled trip hammer is given by Ireland *et al.* (1970) and shown in Figure 2. Another example is the Japanese 'Tombi trigger' shown in Figure 11. These types of hammer give a clean drop, but the height of each drop is subject to driller error. It should be noted that the anvil in Japanese use is only 75mm high by 75mm in diameter, which is very much smaller and therefore lighter than many of those in use in the UK or the USA.

Slip-rope hammers are in widespread use in the USA, Japan, and South America. Figure 12 shows the 'Safety' and 'Donut' hammers, which are used in the United States. In the slip-rope method (Figure 13) the hammer weight is attached to a lifting rope (often manilla or hemp) throughout the execution of the test. The rope passes up from the weight, over a pulley on the rig mast, and down to the operator. The free end of the rope is loosely wound around the power-driven, permanently rotating, steel drum or (or cathead) on the rig. By pulling on the free end of the rope, the rig operator tightens the rope around the cathead, pulling down the rope to lift the weight. The required drop of 30 in is normally marked on the pin or centre rod of the hammer at the start of the test. To deliver the correct energy, the rig operator not only has to lift the weight to the exact distance above the striker plate, but has also to slacken the rope on the cathead quickly and effectively, paying out the rope as well as keeping it free to run as the weight falls. This is difficult, and much depends upon the number of turns of the manilla rope around the cathead (see Serota and Lowther, 1973).

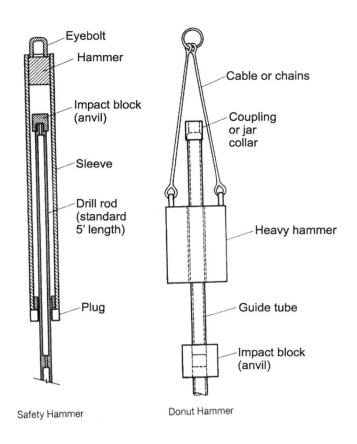

Figure 12 *US SPT hammers (a) Safety hammer (b) Donut hammer (after Seed et al., 1985)*

Figure 13 *Using a cathead to lift the SPT weight in Venezuela in 1987 (note use of about 7/8th turn)*

In Japan, cathead operators typically use only three-quarters of a turn or, perhaps, one and three quarters, and in some cases throw the rope completely off the cathead during the release process (Kovacs and Salamone, 1984). In most other countries the cathead would usually have two or three turns of the rope around it, resulting in significant but erratic energy loss from rope/cathead friction, as the hammer weight falls. Other energy is lost to friction in the pulley bearing on the rig mast.

In Brazil a number of techniques are in use. For one, which has not so far been described, the weight is lifted without powered machinery by one or two men pulling down on a rope passing over a pulley, as was the practice in the USA in the 1920s. The weight is dropped simply by releasing the end of the rope. It does not fall freely because of the inertia of the pulley and friction in the pulley bearing. But the method is preferable to the use of a cathead.

It has been seen that the anvils (striker plates) vary considerably in mass. This is thought to lead to differences in the efficiency of hammers. Further variability comes from the use or omission of wooden cushion materials in the hammer (see Ireland et al., 1970, and Figure 2). Most hammers now in use do not have a hardwood cushion block, and have a different dynamic efficiency from that of the original test. However, pin-guided hammers with wooden cushion blocks are, apparently, still in use in Brazil (NBR 6484-1980). Ireland (1966) comments that the use or not of a wooden cushion block 'is believed to have a major effect on the results of the SPT'.

Rod stiffness varies widely even within countries. For example, Schmertmann (1974) notes that in the USA rods may be as small as DCDMA 'E' or as large as 'N'. ASTM Standard D1586-67 in fact requires that the drill rod must have a stiffness greater than DCDMA 'A' rod (o.d. 41.2mm). In contrast, according to Muromachi et al. (1974), all Japanese rods are between 40.5 and 42.0mm in diameter. Long strings or lighter rods tend to produce higher N values, as a result of energy lost in bending.

Split-spoon geometry also varies to some extent throughout the world. In the USA almost all split-spoon samplers have an enlarged internal diameter of 1-3/8 in (34.9mm) intended to take a sample liner, but they are usually used without these liners (Schmertmann, 1979). This inside clearance improves sample recovery, but leads to a reduced penetration resistance. In many other countries (e.g. South Africa, Japan, Australia, Iran, Brazil, Venezuela) the ASTM spoon is used, which has a ball check-valve in its head. Until 1990, BS 1377 did not specify a split-barrel sampler containing a ball check-valve. In Israel the ASTM split spoon includes a spring catcher to help prevent loss of samples. This probably leads to an increase in penetration resistance. It should be noted that the geometry of the ASTM split spoon is not identical to that shown as the Raymond Concrete Pile Co.'s spoon by Terzaghi and Peck (1948, 1967) nor to that shown by Fletcher (1965). There are minor differences in the cutting shoe details and in the specified port (vent) areas. In Argentina different cutting shoe geometries are used in order to improve sample quality.

Australia, Spain, Turkey and the UK are the only countries known to use the 60° solid cone in gravelly soils. Using the standard shoe in coarse gravels and flinty chalks can give excessively high N values on those occasions when a large particle becomes lodged in the cutting shoe of the split spoon during driving. On the other hand, there is now strong evidence that under some circumstances the cone gives a higher penetration resistance than the shoe (Thorburn, 1986).

The method of SPT testing is far from standardised, even in the USA. ASTM D1586 – 84 requires that once the sampler is resting on the bottom of the hole it should be driven for three increments of penetration of 150mm, or until 100 blows have been applied. The penetration resistance, N, is to be the number of blows for the last 300mm of penetration. If less than 300mm is penetrated, the number of blows and fraction of 300mm penetrated are to be recorded. In its original use the spoon was seated with a few light taps of the hammer, before driving and recording the blows for 1 foot of penetration. This procedure was then modified, omitting the few seating blows and driving the spoon for three 6-in intervals (Schmertmann, 1966). By 1966 there were several methods of carrying out the test; many engineers used the sum of the two lowest blowcounts as the N value (Fletcher, 1965). Others used the sum of the last two blowcounts (Geisser, 1966). Schmertmann (1978) indicates that these differences in practice continue.

4.4 FURTHER DEVELOPMENTS

Developments of the SPT, in terms of the equipment, the way in which it is used, and the purpose to which it is put, are continuing. Notable amongst these are:

1. The use of the SPT as a vertically polarized shear wave source (Stokoe and Abdelrazzak, 1975). Stokoe and his co-worker, in an investigation of the shear moduli of compacted fills, found that the Standard Penetration Test provided seismic energy rich in shear waves, and relatively weak in compression wave energy. Of the three source type they used, they found that the SPT generated the most energy, and therefore led to the most distinguishable vertically polarized shear waves in parallel cross-hole seismic tests. Thus the SPT may be used not only with empirical correlations for G_{max}, but also part of the equipment for direct determination.

2. The recognition that the SPT N value is increased by the presence of large soil particles. Large and very large penetrometers have been used in gravels, in Japan (Tokimatsu, 1988; Yoshida *et al.*, 1988), the USA (Harder and Seed, 1986; Harder, 1988), and Italy (Crova, *et al.*, 1992), instead of the SPT in order to overcome this effect. The use of a larger diameter penetrometer would appear necessary once D_{50} exceeds about 3mm.

3. The introduction of additional measurements. In Japan the practice adopted has sometimes been to measure the SPT penetration for each blow, and in the UK it has been suggested that in granular soils the effects of borehole disturbance can be detected by lengthening the test drive, and measuring the blows beyond the normal 450 mm drive. Recently in Brazil a simple procedure has been added to the standard SPT procedure (Ranzini, 1988). At the end of the test drive, but before the split-spoon is withdrawn from the soil, the driller attaches a torque wrench to the drill rods and measures the maximum torque required to rotate the sampler. This operation not only provides additional information on the soil (Decourt, 1991), but also facilitates the extraction of the split-spoon.

4.5 SYNOPSIS

1. The ASTM standard for the SPT is widely used internationally, although many countries now have their own standard.

2. An International Reference Test Procedure (IRTP) has been approved by the ISSMFE. All SPT tests should, ideally, conform to the essential elements of the IRTP, and these aspects of the SPT should be reported whenever SPT results are published.

3. There are many variations in international SPT practice, which lead to differences in the penetration resistances determined in similar soil types.

4. The main variations lie in

    - the methods of drilling and supporting the hole
    - the hammer mechanisms and rod sizes used.

There are also minor variations in the split-spoon geometry and the method of testing.

A qualitative appraisal of the effects of these different practices on penetration resistance is given in Section 5.

5 The influence of different practices and equipment on penetration resistance

Apart from the soil conditions in which the test is made, the result of a Standard Penetration Test is influenced by three main groups of factors, associated with:

- drilling or boring technique
  - method of hole support
  - method of advancing the hole
  - size of hole

- SPT test equipment
  - hammer design
  - rod size
  - split-spoon design

- test procedure
  - seating drive
  - rate of application of blows.

Differences in drilling technique produce the largest differences in penetration resistance in granular soils. In this respect the driller plays the most important role; the manual dexterity and care which he uses while boring or raising and lowering tools in the hole may have as great an influence on penetration resistance as the soil itself. Test equipment has an important, though generally smaller, influence on penetration resistance. The effects of differences in test procedure are more difficult to evaluate, but are probably similar to those of the test equipment.

The influence of these factors is considered below. For some there is little quantitative information, whilst for others it is difficult to isolate their effects from those of interrelated factors. Figure 14 summarises the information contained in this section of the report.

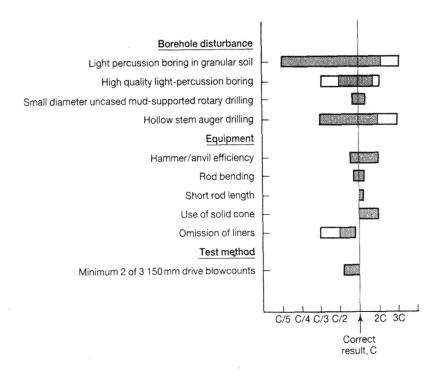

Figure 14 *Summary of effects of different practices and equipment on SPT N value*

5.1 DRILLING OR BORING TECHNIQUE

The factors which may influence penetration resistance are associated with:

- method of hole support
  - hole cased and incompletely filled with water
  - use of bentonite drilling mud
  - casing driven ahead of borehole
  - SPT carried out inside casing

- method of advancing the hole
  - downward – facing flush ports or jetting bits
  - hollow-stem augers
  - shelling and suction
  - use of a claycutter

- size of borehole
  - soil particle size
  - groundwater conditions
  - method of hole support

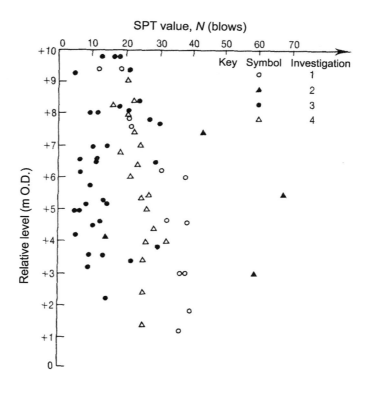

Figure 15 Results of four investigations in the Trent river gravels by light percussion boring (Connor, 1980)

Not all of these factors have been studied. However, the influence of the water level in the borehole relative to groundwater level has been reported for a number of sites in granular soils. For example, Sutherland (1963) reported average N values for seven boreholes in a fine to medium sand where water balance was not controlled, which varied on a borehole to borehole basis from 22 to 83 blows, with an average of 39 blows. Three check boreholes carried out using water balance gave N values of 64 to 94 blows, with an average of 82 blows.

It might be thought, as the provisions of BS 1377:1975 tended to suggest, that this type of effect is only serious in silts and silty fine sands. Common experience demonstrates that this is not so. Figure 15 gives the SPT results from four site investigations, carried out by light percussion drilling with a shell by four different but reputable site investigation contractors in the Trent river gravels. The average N value for the third investigation is about 12 (with almost no discernible increase of penetration resistance with depth), while values from the first investigation are typically 15-20 to 3m, and about 35 below this depth. All the investigations were carried out using 150-200mm diameter equipment; the differences between the four sets of results can be attributed to groundwater, borehole water, and unquantifiable differences in drilling techniques. Nor are such problems restricted to granular soils. The Chalk is equally sensitive to drilling technique, as the results in (Figures 16 to 18) show. Mallard's data (Figures 16 a to d) show that the tightest grouping of SPT results were obtained from small diameter rotary percussive drillholes, and that the scatter of results produced by light percussion drilling included very high penetration resistances, as well as some lower ones (e.g. Contractor C). Clayton (1990a) shows how the type of drilling can significantly change the penetration resistance. Rotary rock-rolled boreholes produced *N* values on average about twice those from a routine light percussion investigation (Figure 17). Montague (1990) (Figure 18) shows how different light percussion rig operators working at the same site could produce significantly different results. Such effects have not, apparently, been reported in other types of weak rocks.

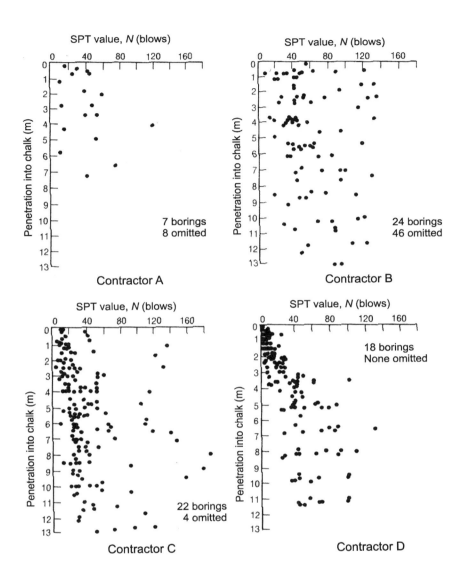

Figure 16   *SPT results from investigations in the Chalk at Littlebrook (Mallard, 1977). Contractors A, B and C used 200-250mm dia. light percussion borings. Contractor D used BX (60mm) rotary percussive holes*

Whilst there are numerous reports of the significant effects of drilling disturbance on penetration resistance in both granular soils and chalk, in clays the SPT appears relatively unaffected. In very soft and soft clays both light percussion boring and auger drilling can cause very significant disturbance ahead of the borehole, because of lack of adequate borehole fluid balance; but the SPT is relatively insensitive to changes in the consistency of such materials (see Table 8, page 74), and is not, therefore, normally considered to give useful data.

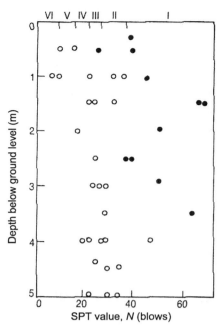

Figure 17 *Comparison of SPT N values for different drilling methods for a site in the Chalk at Leatherhead (Clayton, 1990a)*

It is clear that the combination of driller skill and casing/borehole water level is very important when light percussion boring techniques are used. It can change the penetration resistance of a given soil by a factor of at least five.

The large changes shown by the case records presented above are not solely the result of lack of water balance in the hole. They arise from the interactions between:

1. The use of casing or hollow-stem augers, which (a) concentrates any water inflow into the borehole upwards through the intended SPT test section, encouraging piping, and (b) gives the opportunity for the test to be carried out in confined conditions within the casing or auger if the hole is not properly cleaned. These effects are avoided in uncased, mud-supported boreholes.

2. The use of a tight-fitting shell, which increases the upward hydraulic gradient in the soil below the hole when it is withdrawn. BS 1377:1990 no longer permits the use of any drilling tools with a diameter greater than 90% of the internal diameter of the casing, but it is doubtful if this is sufficient to avoid disturbance.

3. The Borehole diameter. In experiments, Reidel (1929) observed distortions in the soil to depths of approximately three times the hole diameter when soil moved upwards, and it is likely that a greater depth is disturbed when piping takes place.

4. The changing levels of water or mud in the borehole, when the groundwater level is high. Even minor changes, such as a drop in the borehole water level as a drilling tool is pulled to the surface, can be sufficient to induce piping in fine or silty sands (Begemann and de Leeuw, 1979).

Figure 18 *Comparison of SPT(C) results obtained by two different ground*

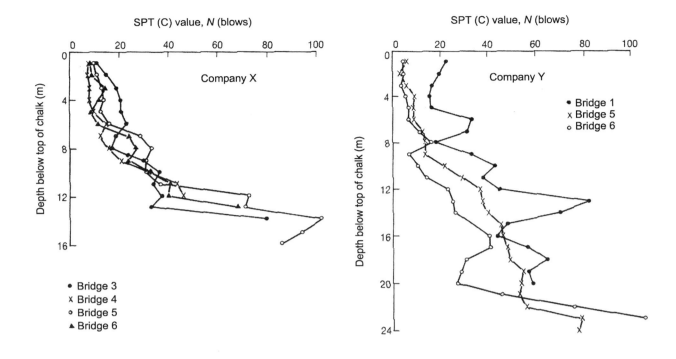

investigation companies, using similar light percussion equipment at the same site in chalk (after Montague, 1990)

Borehole disturbance in granular soils can be reduced by attention to any one of these factors. An example of this is shown by data on the influence of borehole diameter on penetration resistance. Lake (1975) reported reductions 'of the order of 25 – 50% in moving from 125 – 200 mm diameter' light percussion borings. Figure 19 shows SPT results from Cairo, which contrasts values obtained from 76-mm diameter cased rotary water-flush boreholes with those from holes formed using a 200-mm diameter light percussion shell in casing (Greenwood, 1986). Similarly, Jardine (1986, 1989) reported the effect of borehole diameter on penetration resistance, also for sites in Cairo, and showed that the penetration resistance in 150 mm boreholes was typically 30% more than in 200 mm boreholes. On the basis of very limited evidence, Skempton (1986) has suggested correction factors for borehole diameter shown in Table 3.

Table 3 *Approximate borehole diameter corrections proposed by Skempton (1986)*

Borehole diameter (mm)	Correction factor C_d
65-115	1.00
150	1.05
200	1.15

$N_{corrected} = N_{measured} \cdot C_d$

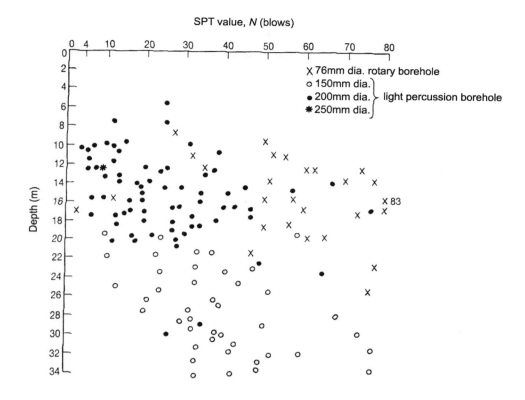

Figure 19   *Influence of borehole diameter and drilling method at Cairo*

It is doubtful if these differences would be so great if comparisons were made between uncased, mud-supported boreholes of different diameters. As early as 1957, Palmer and Stuart noted that upward seepage into the borehole could cause loosening 'of sand or even of gravel' which might falsely lower the penetration resistance. This upflow is concentrated by the use of casing, giving higher hydraulic gradients and a greater chance of boiling. Mallard (1983) suggests that even careful drilling with a full reverse head of bentonite of 6m above ground water level is insufficient to reduce the scatter of penetration resistance values in large-diameter (150 - 300mm) shelled and cased boreholes. In contrast, the use of mud without casing in rotary holes with a diameter of less than 98mm produces a remarkable reduction in the scatter of results (Figure 20).

Although the idea is logical, there appears to be no evidence in terms of comparative N values to support the view that the use of an undersize shell has any effect in reducing the loosening of granular soils. Because the rate of boring with undersized shells is slow, they are rarely used. However, Nixon (1954) shows photographs of sand samples with undistorted horizontal layering, taken with the Bishop Sand Sampler, which demonstrate that, with good technique, disturbance can be minimised. The method involved:

- the use of an undersize shell
- continuous feeding of water to the borehole
- pulling the shell slowly enough to prevent the borehole water level from rising.

Methods of boring or drilling which add water or flush fluid to the borehole (such as wash-boring or rotary drilling) have the advantage that by keeping the borehole full of fluid, they ensure an outflow from the bottom of the hole during the boring and cleaning process. Methods which tend to remove fluid, such as light percussion boring or hollow-stem auger boring, can lead to significant loosening. Hollow-stem auger boring may either loosen the soil (for example, when the central plug is removed) or densify it (by downthrust of the augers on the test section). Schmertmann (1978) estimates that the use of water and casing (as opposed to drilling mud without casing) may reduce penetration resistance by 50%, and that the use of hollow-stem augers may lead to further changes equivalent to either doubling or halving the penetration resistance.

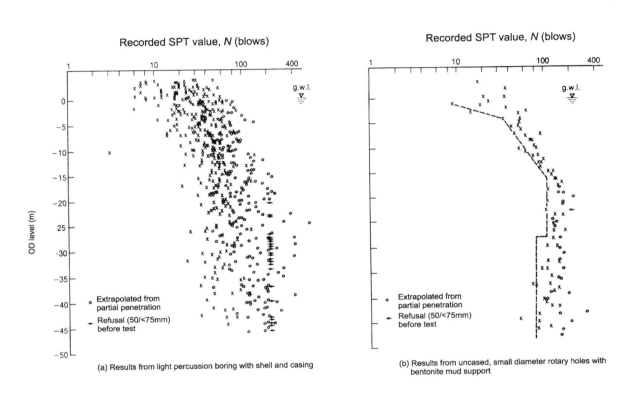

Figure 20 *Comparison between SPT results in Norwich Crag sands, Suffolk, from light percussion and rotary borings (after Mallard, 1983), (a) with shell and casing (b) uncased, bentonite mud supported small diameter rotary borings*

There appear to be no published case records to indicate the influence either of driving the casing ahead of the base of the hole, or of subsequently carrying out the SPT when there is soil within the casing. Qualitative observations made by the writer indicate that these practices can increase the penetration resistance by at least 200%. In some countries drilling practices are so poorly controlled that progress is made by advancing the casing by several metres, carrying out the SPT within the casing, and then cleaning out; but in other cases drillers may induce a similar effect when they inadvertently carry out the test within the casing after soil has flowed up inside it, or when sand and gravel settle out of the borehole fluid. An illustration of both the technical and financial implications of this is given in a case history by Naismith (1986). In the offshore investigation for a piled wharf, problems were reported of gravelly sand, which was encountered to a depth in excess of 50 m, heaving into the casing during washboring. The SPT values indicated that the gravelly sand subsoil was in a 'compact to dense' state, and on this basis the geotechnical designers recommended a pile length of 15 m. During construction the piles drove easily, and there was no sign of increased bearing resistance at the design elevation.

After further driving, and a pile load test, a satisfactory pile length was established. It was approximately double the original pile length. Since there were insufficient piles on site the construction programme was delayed for a year. Both the owner and the main-works contractor claimed against the geotechnical firm.

5.2 TEST EQUIPMENT

Two techniques have been widely used to assess the influence of test equipment on penetration resistance. The first, already noted in previous sections, compares N values obtained from the same site, but using different equipment or a different drilling technique. The natural variability of the soil may invalidate such comparisons, and at least makes them uncertain. The second technique, which has been used to look at hammer system variability, makes comparisons between the energy delivered to the SPT rods. It has been demonstrated by Schmertmann and Palacios (1979) that, at least up to $N = 50$, penetration resistance varies inversely with the energy, E, transmitted to the split spoon by the rods during the first compressive wave pulse, provided the pulse is of sufficient duration (see Appendix 2). The correction factor for penetration resistance, once the energy has been measured, is:

N (measured) / N (standard) = E (standard) / E (measured)

This relationship was first suggested by Palmer and Stuart in 1957.

Recent international interest in the measurement of SPT energy probably stems from the relative simplicity of this approach, and the fact that energy measurement has become possible using modern instrumentation. ASTM Standard (ASTM D 4633-86) provides guidance on energy measurement (see Appendix 2 and Clayton, 1990b).

In Sections 3.3 and 4.3 attention is drawn to the following aspects of equipment design:

- hammer design
  - energy delivered to anvil by automatic trip hammers, hand controlled trip hammers, slip-rope hammers, and hand-lifted hammers
  - influence of anvil mass
  - influence of wooden cushion blocks

- rods
  - rod whip in flexible and bent rods
  - length of rod string
  - rod weight and stiffness
  - energy losses in loose couplings

- split-spoon design
  - use of catchers
  - lack of liners in US larger i.d. samplers
  - vent area and ball-check valve details
  - penetration resistance of cone compared with shoe.

Few SPT hammers deliver energy to the rods equivalent to that of a 63.5kg weight falling freely through 762mm (Serota and Lowther, 1973; Schmertmann and Palacios, 1979; Kovacs and Salamone, 1982; Kovacs et al., 1982; Riggs et al., 1983; Seed et al., 1985; Skempton, 1986; Clayton, 1990b). Skempton has suggested that for the purpose of comparing penetration resistances from different systems a standard rod energy ratio of 60% may be convenient, since this value is in the middle of the range of energies measured for different hammer systems. Table 4 shows the variation in N value from that produced with this proposed standard energy, for different hammer systems. The symbol, N_{60}, is used to denote N values corrected to 60% of the theoretical free-fall hammer energy.

Extensive testing of British automatic trip hammers has not yet been undertaken, but data from Skempton (1986) suggest that the use of a heavy (approximately 16-19kg) anvil as compared with a light (2kg) Japanese anvil would result in a rod energy loss of about 23%. Analyses using the wave equation (Smith, 1962; McLean et al., 1975) suggest that the equivalent energy delivered by a hammer with a 20kg anvil will be only 90% of that with a 2kg anvil. Limited energy measurements made by the writer (Clayton, 1990(b)), using a Dando automatic trip hammer, gave rod energy measurements of 70-75% of the free-fall hammer energy. A figure of 70% has been used in Table 4, giving an energy correction (to 60% of the free-fall energy) of 1.16 – 1.25.

Table 4   *Estimated influence of hammer type on SPT N value*

Country	Hammer type	Release mechanism	% variation in N from suggested standard
UK	automatic	trip	– 15
USA	safety	2 turns on cathead	0 to +10
	donut	2 turns on cathead	+35
Japan	donut	Tombi trigger cathead	– 25
	donut	2 turns + special release	– 10
China	automatic	trip	0
	donut	hand dropped	+10
	donut	cathead	+20
Argentina	donut	cathead	+35

Note: based upon suggested standard of 60% rod energy ratio (Skempton, 1986), and data from Seed *et al.* (1985), Skempton (1986) and Clayton (1990b) (see also Appendix 2).

The greatest difference due to hammer efficiency shown in Table 4 is 35%, which is minor when compared with the differences induced by borehole disturbance in granular soils discussed in Section 5.1 (Figure 14). But the figures in Table 4 are based upon averages of the data available for each hammer type. Reference to detailed, blow by blow, data for hammer systems show how inconsistently some deliver their energy. Slip-rope hammers, for example, are particularly poor, with energies from as little as 30% to as much as 85% of the free-fall hammer energy (Schmertmann and Palacios, 1979). These limits give correction factors of –30% and +100%. Automatic trip hammers are likely to be much more consistent, with data obtained by the writer giving 95% confidence limits of the order of a few percent.

As far as is known, none of the energy measurements used as the basis for Table 4 were made on hammers with a hardwood cushion block. There are currently no published comparisons of the energy transmitted with and without a cushion block, but wave equation analysis suggests that the effect of a hardwood cushion block might be to increase blow counts by as much as 30%.

It has been thought for many years that rod stiffness and rod weight might significantly influence dynamic penetration resistance. Flexible rods in long strings would be expected to absorb energy in bending, while heavy rods might cause significant inertial losses.

With regard to the influence of rod stiffness, some of the smallest rods used for the SPT are the 40.5mm diameter rods specified in JIS 1219. Koreeda et al., (1980) compared the results from duplicate borings in different soils, as shown in Figure 21, and concluded that for all practical purposes there are no significant differences between the results when using 40.5mm and 50mm diameter rods. If anything, the 50mm rods led to slightly higher N values in hard clays below 15m depth. The effect of rod length on the efficiency of energy transmission to the SPT split spoon has also been investigated by Uto et al. (1973, 1975) and Matsumoto and Matsubara (1982). As Figure 22 shows, the efficiency of energy transmission is high even using 40.5mm diameter rods, laterally unsupported, at depths of more than 15m. The Japanese have therefore concluded that long strings of 40.5mm diameter rods do not significantly influence penetration resistance. Work in the USA by Brown (1977) showed no measurable difference in penetration resistance obtained with A and N rods, which seems to suggest that inertial effects are small, at least up to this rod size.

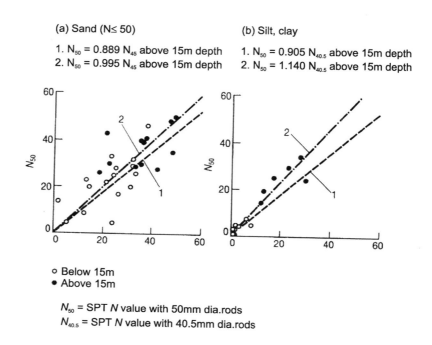

Figure 21 *Influence of rod diameter on penetration resistance (from Shioi et al., 1981) (a) sand N=50, (b) silt, clay*

The effect of short rod lengths on energy transfer is less certain. It can be theoretically predicted that in soils with a low penetration resistance (less than $N = 50$) the energy transmitted down the rod in the first compressive pulse of force will be reduced, as a result of a reflected tensile wave (see Appendix 2). On this basis Skempton (1986) suggested that a correction should be applied for tests carried out with rod lengths of less than about 10 m. However, direct energy measurements and observations of the ratio of undrained strength to uncorrected penetration resistance indicate that such a correction is not required. Figure 23 (a) shows that the measured energy is not reduced for depths as little as 4 m, and Figure 23 (b) shows that the ratio cu (remoulded)/*N* is not altered even for rod lengths as little as 1.5 m.

No test results are known concerning the influence of core catchers or vent arrangements on penetration resistance. Data exist, however, to allow the influence of the 60-degree cone, and lack of liners in US practice, to be assessed.

Palmer and Stuart (1957) introduced the solid 60-degree cone-ended SPT, and argued, from limited tests, that the penetration resistance of this tool was effectively the same as with the split spoon. Because the cone has a longer life than the shoe in gravel, and to avoid obtaining unrealistically high N values if a large particle were to become lodged in the standard show, the British site investigation industry has widely used this modification. However, data from Thorburn (1986) show that this change may, in granular soil, lead to an increase in penetration resistance of as much as 100% (Figure 24). In chalk this effect has also been observed at a number of sites (e.g. Figure 25), but at others no effect is observed (Figure 26).

The effect of the apparently common US practice of omitting liners from split-spoon samplers with an enlarged inside diameter has been evaluated by Seed *et al.* (1985). As Figure 27 shows, this leads to a decrease in penetration resistance of about 10-30%, as compared with the result when using a standard 35mm (1-3/8in) i.d. sampler.

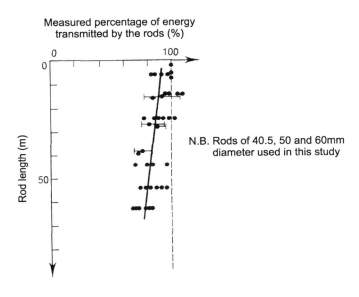

Figure 22   *Effect of long rod lengths on transmitted energy (after Uto et al., 1973, 1975; Matsumoto and Matsubara, 1982)*

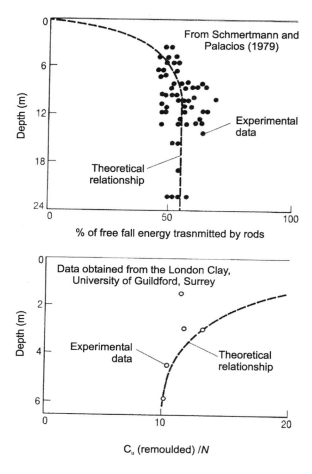

Figure 23 Effect of short rods on energy transmission and c_u/N ratio

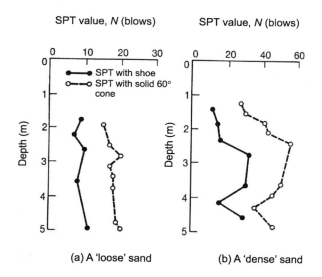

Figure 24 Comparison of penetration resistance with the standard SPT shoe (SPT) and the 60° sand cone (SPT(C)) in sands (after Thorburn, 1986)

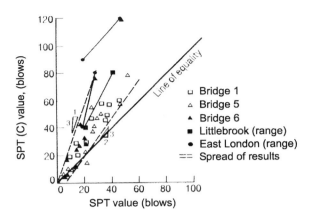

Figure 25 *Comparison of SPT (i.e. open shoe) and SPT(C) (i.e. 60° cone) N values in chalk (after Montague, 1990)*

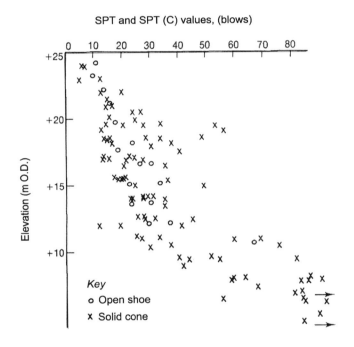

Figure 26 *Comparison of SPT and SPT(C) N values for a site in the Chalk at Brighton*

5.3 TEST METHOD

The major influences of test method are:

(a) the way in which the split-spoon is seated at the base of the borehole

(b) the datum from which penetration is measured, and

(c) the way in which penetration resistance is measured.

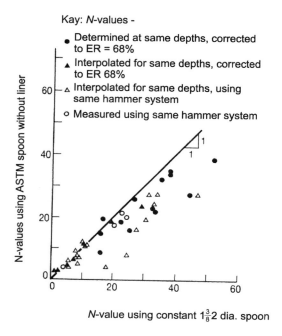

Figure 27   *Effect of omission of split-spoon liners on penetration resistance (after Seed et al., 1985)*

In very loose granular deposits, or in very soft clays or soliflucted chalk, the self-weight of the rods and hammer assembly will be sufficient to push the penetrometer some distance into the ground at the base of the borehole. The current British Standard gives good guidance on what should be done, but in many instances the level of the bottom of the borehole is not determined and it is assumed that the penetrometer comes to rest exactly at the bottom of the hole. Penetration resistance will then be over-estimated by an unknown quantity. In other instances, the split-spoon is given a few light taps, in order to seat it, before the drive is started. This will lead to a greater over-estimate of N.

Penetration resistance is measured in many ways around the world. In some instances it is taken as the sum of the two smallest of three blowcounts for 150 mm. At the opposite extreme, in some Japanese tests, penetration has been measured for every blow. Where a full drive is achieved, and the *least* two blowcounts are added to give the N value, it can be estimated from Schmertmann's data (Schmertmann, 1979) that this determination of penetration resistance may be about 30% less than using the final two 150-mm blow counts.

A more serious lack of standardisation occurs when the test is used in weak rock. The test drive may be terminated after a variable number of blows, and the effects of boring and drilling will be larger the smaller is the length of the test drive. This effect has not, apparently, been reported in the literature, but it is clear that the longest possible test drive is desirable and that the split-spoon penetration should be recorded at more frequent intervals than is normal in soils. When testing in rock it may be necessary to vary the split-spoon geometry, as is done in Venezuela (Hiedra-Cobo, 1987), to avoid damage to it. The current British Standard does not give any guidance on the extrapolation of the resistance measured in a short drive.

5.4 SYNOPSIS

1. Figure 14 summarises the effects of drilling, test equipment and test procedure on penetration resistance. The major factors are:

   - in granular soils, and chalk, the disturbance caused by drilling
   - in sands, the use of a solid cone in place of the spoon
   - the omission of liners in the spoon
   - the efficiency of the hammer.

   Of these the disturbance caused by drilling has the greatest effect.

2. Drilling disturbance may be particularly large in uncemented granular deposits, and in chalk. Most, but not all, factors are likely to lead to blow counts which are too low. Exceptions can occur when rod energy is low or the bottom of the casing is below the test section.

3. The use of the SPT cone should be discouraged, except where essential, such as in coarse granular soils. When the intention is to measure N values, standard 35mm i.d. split spoons should be used, i.e. without liners.

4. Hammer efficiency may be checked using energy measurement techniques (see Appendix 2). In the UK it is thought that the SPT is well standardised, giving about 70-75% of the free-fall hammer energy. SPT equipment used abroad should be carefully evaluated, because the energy delivered by the hammer may be quite variable and, on average, different to that given by UK automatic trip hammers.

5. Other influences, such as rod bending, are thought to be of generally minor importance. But such factors cannot always be ignored, and may be considered important under particular circumstances.

6. In clays, drilling disturbance is generally small, and the results of the SPT are therefore relatively reliable.

7. SPT results are used in empirical calculations, and the test is therefore the basis upon which experience is transferred from site to site, and from engineer to engineer. This Section has shown that, under certain circumstances, the test may be greatly affected by factors other than the ground. These factors should be taken into account when applying the methods of calculation given in Sections 8 and 9.

6 The influence of ground conditions on penetration resistance

In Section 5 it was seen that the method of forming a borehole, and the equipment used for carrying out an SPT, have a major influence on the test result. In this Section the influence of the ground is examined.

The SPT spoon, 50mm in diameter, is forced into the ground by short pulses of hammer energy passing down the rods. Ground resistance is provided by friction on the outside and inside of the spoon, and by end-bearing. Thus the N value results from a test on a relatively small volume of ground, that is performed dynamically and intermittently, to failure, and beyond, in conditions which may be either undrained or partially drained.

The SPT should be viewed in the context of the range of *in-situ* tests available for the characterisation of the ground. For comparison, some of the tests used in the UK are compared and contrasted in Table 5, and are discussed below.

Table 5 *Comparison of in-situ tests*

TEST TYPE	large range of design use	works in many types of ground	low cost	simple to carry out	easy interpretation	uniform stress paths	uniform strain levels	well-defined drainage conditions	data obtained pre-yield	data obtained post-yield
pressure meter	●	○		●	●	○				
plate test		○		●	○			○		○
vane test		○			●				●	●
dilatometer		○	●	●			○	●	○	
CPT		○		○		●	○	●		●
SPT					○	●	●	●1	●	

○ test has some good features in some situations
● most positive features of the test

Ideally there should be a direct analytical method by which the measurements made during an *in-situ* test can be converted to the geotechnical parameters required as an end-result (for example, to allow the undrained shear strength of a soil to be determined from torque measurements made during a vane test). As soil behaviour depends both upon the stress path and the strain levels to which it is subjected, an ideal test would impose uniform stress paths and strain levels on all the elements of the ground around it (regardless of their position, since measurements can only be made at the boundary of the test apparatus). In addition, all the elements under load should remain either completely undrained or pore pressures should be allowed to dissipate fully.

Furthermore, an *in-situ* test should apply a sense, direction, scale and rate of loading which is relevant to the geotechnical phenomenon (e.g. soil compression) for which design parameters are needed. In principle, not only should the test model follow closely the direction of stress change (i.e. horizontal or vertical) which will be applied by the full-scale structure, it should also make changes in the correct sense, for example unloading to simulate the influence of the movement of a retaining structure towards an excavation, or loading to provide parameters for foundation analysis. Apart from these considerations, which can only partly be satisfied by any given *in-situ* test, it is also important that the size of the apparatus is large so that the soil under test contains representative particles, fissuring, jointing or drainage fabric. Only then can measurements reflect the strength, compressibility or drainage characteristics of the soil or rock mass rather than its component parts.

No *in-situ* test yet devised can satisfy all of these requirements. For example, the results of self-boring pressuremeter tests are particularly amenable to analysis in certain soil conditions, such as overconsolidated clay, but such a test cannot apply a uniform strain to all the elements of soil it affects. Thus measurements of stiffness are at best an average of the behaviour of the ground at different strain levels. And without the ability to bring a sample of the test section to the ground surface, it can never be guaranteed that the test results are not invalidated by the presence of fabric which might (for example) allow rapid drainage of the test section, or make the size of the instrument too small for the measurement of mass compressibility of a soft rock.

Experience shows that *in-situ* testing methods become used by practising engineers not because of the elegance of analytical methods associated with them, but because:

1. The methods can be guaranteed to work in given soil conditions.
2. The tests provide parameters that cannot more easily or more cheaply be obtained in another way.

In these respects tests such as the pressuremeter, the plate test and the static cone are often unattractive for routine investigations in British soil conditions. The SPT, which gives very little information, is not logically interpretable in terms of the geotechnical parameters normally required for design, and is poorly standardised and crude, yet remains very attractive because it is sure to work in almost any ground conditions, is relatively cheap, and normally yields a sample of the ground in which the test is carried out. It is probably the most widely used soil test in the UK.

6.1 THE SPT IN GRANULAR SOIL

Setting aside the influences of boring discussed in Section 5, there are a very great number of factors known to influence the resistance of a dynamic penetrometer in granular soil (Table 6). The SPT causes dynamic failure of the soil, and so penetration resistance should be a function of the effective angle of friction and effective stresses operational at the time of the test. But the speed of the test means that excess pore pressures are generated, even in uniform clean sands when the ground is relatively permeable (Clayton and Dikran, 1982). The fact that conditions in the test section are intermediate between drained and undrained makes interpretation impossible even if it is assumed that the operational total stresses during testing are the same as those which acted upon the soil before the test.

With respect to the influence of excess pore water pressures, Terzaghi and Peck (1948) suggested that, in silty soils when $N > 15$, there would be an increase in penetration resistance. On this basis they argued that penetration resistance should be reduced to

$$N = 15 + 0.5 (N-15)$$

Perhaps because of criticisms by Meyerhof (1956), this correction was not reiterated in the 1967 edition of Terzaghi and Peck. But, more recently, Burland and Burbidge (1985) found that their settlement predictions were marginally improved by its use, suggesting that excess pore pressures do influence penetration resistance.

Table 6 *Influence of granular soil properties on dynamic penetration resistance*

Factor	Influence	Reference
Void Ratio	Decreasing void ratio increased penetration resistance	Terzaghi & Peck (1948), Cribbs & Holz (1957), Holubec & D'Appolonia (1973), Marcuson & Bieganousky (1977)
Average particle size	Increased particle size gives increased penetration resistance	Schultze & Menzenbachi (1961), DIN4094, Clayton & Dikran (1982), Skempton (1986)
Coefficient of uniformity	Uniform soils exhibit lower penetration resistance	DIN4094, Part 2
Porewater pressure	Dense fine soils dilate to increase penetration resistance: very loose fine soils may liquefy during testing	Terzaghi & Peck (1948), Bazaraa (1967), de Mello (1971), Rodin *et al.* (1974), Clayton & Dikran (1982)
Particle angularity	Increased angularity gives increased penetration resistance	Holubec & D'Appolonia (1973), DIN4094
Cementation	Cementation increases penetration resistance	DIN4094, Part 2
Current stress levels	Increased vertical stress gives increased penetration resistance: increased horizontal stresses increase penetration resistance	Zolkov & Wiseman (1965), de Mello (1971), Dikran (1983), Clayton *et al.* (1985)
Age	Increasing age leads to increased penetration resistance	Skempton (1986), Barton *et al.* (1989)

The strength of a granular soil depends upon many factors. The effective angle of friction is a function of stress level, grain size distribution, particle angularity, void ratio (normally expressed in terms of relative density), and to a certain extent the mode of failure. And in many granular soils, strength is increased as a result of particle 'locking', or inter-granular cement, or both. The action of penetrating the soil leads to shearing, dilatancy or collapse, and consequently to changes in total stress.

These stress changes are only partially compensated by changes in pore pressure. Since only one measurement (i.e. the penetration resistance, N) results from the test, any method of interpretation must make the assumption that for all soil conditions N is unaffected by all but a single soil parameter. This type of interpretation is clearly unrealistic, but has been widely attempted.

It is known (Meyerhof, 1957) that, for uncemented quartz medium-grained normally consolidated sand:

- penetration resistance increases approximately linearly with depth, and therefore with vertical effective stress for a constant density

- at constant vertical effective stress, N increases approximately as the square of relative density

- for a given relative density and vertical effective pressure, N is greater for coarser soils.

Assuming that dynamic penetration resistance is proportional to the mean effective stress on the test section immediately prior to the test (Clayton et al., 1985), Skempton (1986) suggested that

$$N = D_r^2 \left(a + b\, C_{oc} \frac{\sigma_v'}{100} \right)$$

where N is the SPT resistance

D_r is the relative density (per unit, i.e. expressed as a decimal)

α is a material dependent factor, found by Skempton to be in the range 17 – 46 for several *in-situ* Japanese sands

b is a material dependent factor found to vary from 17 – 28

C_{oc} is a factor to allow for the increase in penetration resistance as a result of overconsolidation

σ_v' is vertical effective stress in kPa.

Over-consolidation leads to an increase in the effective horizontal stress in a deposit, and in increasing the mean effective stress ($\sigma' = 1/3\,(\sigma_v' + 2.\sigma_h')$) produces an increase in penetration resistance. For an overconsolidation ratio (OCR) of unity, C_{oc} is unity; as the OCR increases to 3 and then to 10, C_{oc} rises to 1.4 and then to 2.3.

At any given site, σ_v' can normally be estimated with reasonable accuracy. The effect of overburden pressure can then be removed, if the sand is assumed to be normally consolidated, by normalising the penetration resistance to its equivalent value at an effective overburden pressure of 100 kPa (1kg/cm²), using a depth correction factor C_N where

$$C_N = \frac{N_1}{N_{\sigma_v}} = \frac{D_r^2 (a + b)}{D_r^2 \left(a + b \dfrac{\sigma_v'}{100} \right)} = \frac{a/b + 1}{(a/b + \sigma_v'/100)}$$

Data used by Skempton (1986) gave α/b values ranging approximately from 0.6 to 1.4. If these limits are taken as valid for all granular soils, a reasonably small range of correction factor arises, and previous factors such as those put forward by Thorburn (1963), Tomlinson (1965), Peck *et al.* (1974) and Seed *et al.* (1983) are found to be in reasonable agreement. Suggested correction factors, taking into account both effective overburden factor and *in situ* K_o are given in Figures 28 a and b.

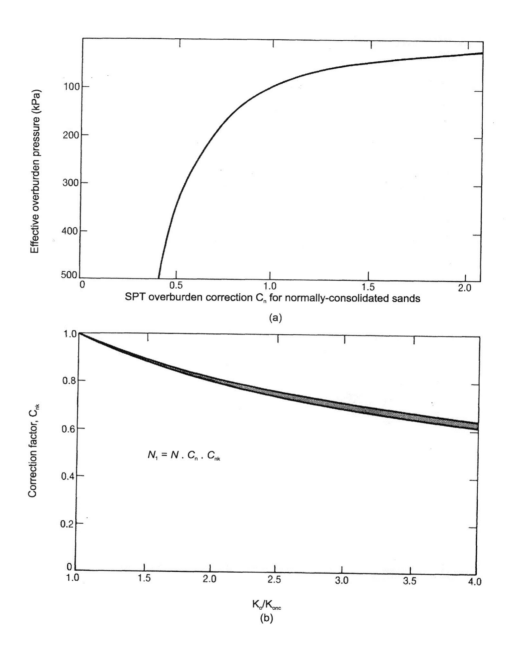

Figure 28 *Corrections for overburden pressure and overconsolidation (a) correction factor for overburden pressure (after Liao and Whitman, 1985; Jamiolkowski et al., 1985) (b) correction factor for increase in mean stress due to overconsolidation (after Liao and Whitman, 1985: Tokimatsu, 1988)*

Even after correction for the effects of effective overburden pressure, penetration resistance cannot readily be converted to engineering parameters, since it remains affected by density (e.g. relative density), particle shape and angularity, particle size distribution (which influence angle of internal friction), over-consolidation and cementing.

The influence of density may be assessed for relatively recent normally consolidated sand deposits, where it may be reasonable to assume that

$$\frac{(N_1)_{60}}{D_r^2} = 60$$

where $(N_1)_{60}$ is the blow count normalised to an effective overburden pressure of 100 kPa (1 ton/ft^2) and corrected to 60% of free full energy, in order to obtain some estimate of relative density (Skempton, 1986). But the values of relative density arising from this should be viewed with some caution since even the limited database used shows $(N_1)_{60} / D_r^2$ values varying from 33 – 84.

At present, in the UK, at least two different systems are in use to obtain the density descriptors, used in sample description for granular soils, from SPT results. In the first case the density is described on the basis of the uncorrected N values. Practitioners using this system regard the use of the descriptors merely as shorthand, conveying only the typical penetration resistance of a stratum. This system (for example, Clayton et al. 1983) is as follows:

Classification	N (blows/foot)
Very loose	0–4
Loose	4–10
Medium	10–30
Dense	30–50
Dense	> 50
Very dense	

In the second case, the values of N are corrected for overburden pressure before the classification system is applied. The use of $(N_1)_{60} / D_r^2$ of 60 gives rise to a relative density classification (after Skempton, 1986) similar to that proposed by Terzaghi and Peck (1948), for sand:

Classification	D_r (%)	$(N_1)_{60}$ (blows/300mm)
Very loose	0 – 15	0 – 3
Loose	15 – 35	3 – 8
Medium	35 – 65	8 – 25
Dense	65 – 85	25 – 42
Very dense	85 – 100	42 – 58

This form of relative density classification is to be preferred to that which obtains relative density directly from uncorrected penetration resistance since it makes clear the effect of bulk excavation on site (see Meigh and Nixon, 1961; Zolkov and Wiseman, 1965). Figure 29 shows how SPT tests carried out after general excavation on a site for a nuclear power station were more than halved near to the ground surface after excavation, as a result of effective stress reductions. It is clearly dangerous to describe a dense granular soil, which gives a small N value because of the low effective stress level applied to it, as very loose. Yet this is what is implied by some current practice.

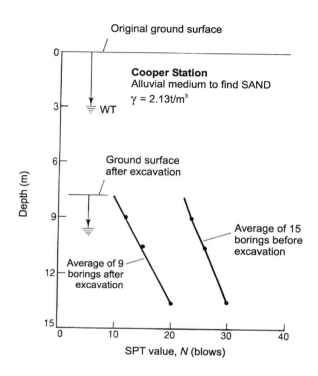

Figure 29 *Effect of decrease of vertical effective stress, due to excavation, on SPT N values (after Lacroix and Horn, 1973)*

While the effects of stress level and relative density on penetration resistance are quite well understood, at least for sands, the influences of other factors are not. For example, it has been postulated (e.g. de Mello, 1971) that the mobilised effective angle of friction of a granular soil has a major influence on penetration resistance. The well known relationship between N and ϕ, suggested by Peck *et al.* (1953), indicates that penetration resistance increases almost linearly with effective angle of friction (see Figure 35 a). But Stroud (1989) has shown that the influence of ϕ is more complex; at low relative densities penetration resistance appears unrelated to angle of friction, with only a poor correlation (because of the unknown influence of overconsolidation ratio, OCR) developing at higher densities.

There is strong evidence that the SPT N value is affected by the particle size of the soil into which it is driven. Japanese SPT data (Figure 30 from Tokimatsu, 1988) shows that once the mean particle size (D_{50}) of the soil exceeds about 0.5mm, penetration resistance climbs rapidly. Whether this effect is genuinely caused by a change in particle size, or whether the increasing particle size is linked to a change in other grain characteristics which give an increase in the effective angle of friction of the soil, is not known. But clearly correlations for sands cannot necessarily be assumed to be valid for gravels.

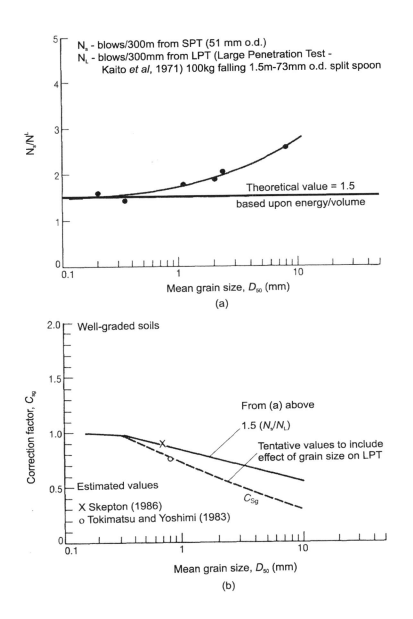

Figure 30 *Effect of grain size on SPT values, (a) effect of mean grain size on Ns/NL ratio (b) tentative grain size correction for SPT*

Another important effect, that of ageing, has been observed for British sands by Barton *et al.* (1989). They suggest that increasing age leads not only to increased density and locking, but also to an increase of penetration resistance above that to be expected solely from that density increase. With $(N_1)_{60}$ values as high as 300 for Jurassic sands, it is apparent that ageing is a further complicating factor in the interpretation of the SPT in granular soils. Other factors of importance, but which as yet have not been investigated, are cementing and mean particle size.

6.2 THE SPT IN COHESIVE SOIL

From the earliest use (for example, Terzaghi and Peck, 1948) it has been appreciated that penetration resistance in cohesive soil is broadly a function of undrained shear strength cu. The precise relationship between undrained shear strength and SPT N is controlled, however, by a number of factors. These include plasticity, sensitivity and fissuring. Equipment factors, such as those described in Section 5, have also undoubtedly contributed to the wide range of the ratio c_u/N reported in the literature. And it is well recognised that undrained shear strength is not an unique soil parameter but depends strongly upon the method by which it is determined (for example, see Wroth, 1984), and the precise orientation of any planes of weakness (i.e. fissures) in the test specimen.

For overconsolidated clays, Stroud (1974) has reported good correlations between N and c_u. (Figure 31). The strength of these correlations results from the standardisation of the SPT in the UK, the relatively small influence of British drilling methods on these types of ground, and the fact that the undrained shear strengths were determined in a single way, using triaxial compression tests on 102mm diameter specimens. It should also be noted that the relationships were developed from comparison of depth profiles of both c_u and N on given sites, rather than comparison of individual results. These depth profiles show that N values are less scattered than undrained shear strength values, and this suggests that the Standard Penetration Test may be a powerful technique for determining undrained strength.

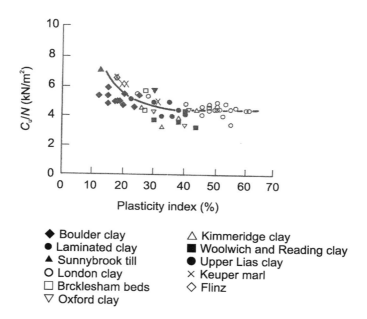

Figure 31 *Correlation between N value and undrained shear strength, cu for insensitive clays (after Stroud, 1974)*

Some comparisons between penetration resistance and undrained shear strength have given a very wide range of c_u/N values. For example, de Mello (1971) shows values with cu/N ratios apparently varying between 0.4 and 20. The contrast between the very tight grouping of results quoted by Stroud (1974) and the wide range of values given by de Mello's data is explained because:

1. Stroud's data relates only to insensitive overconsolidated clays, whereas some of de Mello's data undoubtedly is derived from soft sensitive clays.

2. The undrained shear strengths used by Stroud were obtained from 100 mm diameter unconsolidated undrained triaxial tests (thus reflecting the weakening effect of fissure fabric), whereas de Mello used results from a variety of tests (for example, uniaxial unconfined compression tests), some of which were carried out on very small and possibly highly remoulded specimens.

Sensitivity can be expected to have a significant influence on penetration resistance. Schmertmann (1979) compared the dynamic penetration of the SPT to that of the Dutch cone and concluded that, for clays, at least 70% of the soil resistance can be derived from side shear, the remainder coming from end-bearing capacity. Since end-bearing capacity is determined by undisturbed undrained shear strength, and side shear by remoulded strength, a modest sensitivity of 10 will increase a c_u/N value from 5 (for an insensitive clay) to 13.5 (if the clay is sensitive).

An estimate of the influence of fissuring on the c_u/N ratio can be obtained for the insensitive London Clay by examining the value of this ratio when remoulded samples are used to determine undrained shear strength. Figure 23b gives c_u/N averaging about 11, or approximately twice that obtained by Stroud for fissured London Clay.

The compressibility of a clay cannot be expected to have any significant influence on SPT penetration resistance. The SPT produced dynamic failure conditions, and therefore compressibility/penetration resistance correlations will depend upon the broad relationship between the undrained strength of the material and its stiffness, which occurs as a result of the influence of void ratio upon the two variables.

6.3 WEAK AND WEATHERED ROCKS

The SPT provides one of the very few techniques by which information can be obtained on the mass properties of weak and weathered rocks. Some of the factors known to influence penetration resistance in this type of ground are:

- strength of the intact rock
- porosity of the intact rock
- spacing of joints
- aperture and tightness of joints
- the presence of hard inclusions (such as flint or chert).

In addition equipment and drilling factors may be significant (see Section 5) as may the precise method of the test, particularly with respect to the way the seating drive is carried out. In the following section the penetration resistance of chalk is used to demonstrate some of these points.

When joints are widely spaced and tight, such as in less-weathered weak rocks, the resistance of the rock mass to penetration is a function of porosity and intact strength. A high porosity allows the fractured rock to be pushed aside; a low strength makes it easier for the split spoon to fracture the rock (Meigh, 1980; Leach and Thompson, 1979). For the chalk, the majority of which has a relatively constant chemistry, the wide range of intact strength is reflected by changes in porosity, density and saturated moisture content. Even relatively modest changes in saturated moisture content have been shown to have a large influence on penetration resistance, as Figure 32 shows.

Even when unweathered, the weakest chalks to be found in the UK (saturated moisture content 35-40%) have an SPT *N* value of only about 6-15 blows/300 mm. Relatively unweathered hard chalks (saturated moisture content 14%) have been found to have *N* values of 100-300 blows/300 mm (Woodland *et al.*, 1989), and it is probable that these values are underestimates as they were obtained by extrapolating from penetrations measured for 50 blows.

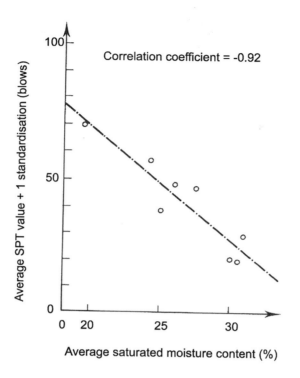

Figure 32 *Correlation between SPT N value and average saturated moisture content of intact chalk (Clayton, 1978)*

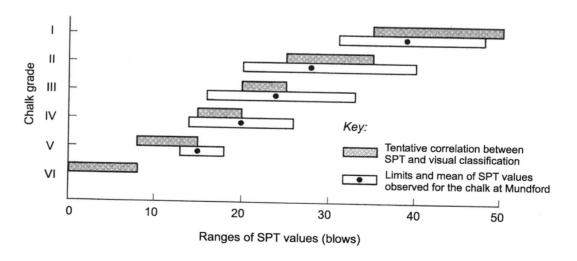

Figure 33 *Correlations between SPT N value and chalk visual weathering grade (after Wakeling, 1970)*

As fractures become more frequent, penetration resistance is reduced. Penetration occurs either by pushing blocks aside, when the split spoon runs down a vertical joint, or more frequently when blocks are split (Thompson, 1980; Webb, 1970). The energy necessary to split an intact block of weak rock will depend upon its size and shape as well as its intact strength. For the chalk at Mundford, the variations of SPT N were found to be related to fracture spacing and aperture (Grades V to II, Figure 33) by Wakeling (1970), with a further influence from higher density and strength in the highest quality (Grade I) chalk.

The presence of flint or chert in the SPT test section can have a significant influence on the result of the test. It is common practice to examine closely the detailed results of SPT in chalk, to look for and eliminate high penetration resistances which might be thought to result from striking flint.

Unfortunately the dependence of penetration resistance on a number of independent factors makes precise interpretation of the test impossible. For example, Dennehy (1975) showed that for most values of penetration resistance the visual weathering grade (Ward et al., 1968) could lie within at least four grades (Figure 34) and therefore that the N value could not be used to determine visual weathering grade.

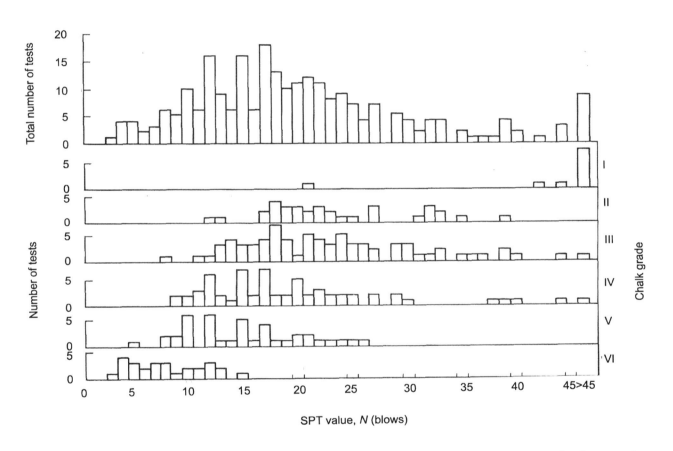

Figure 34   Lack of correlation between SPT N value and visual for weathering grade in the Chalk of Hampshire (after Dennehy, 1975)

In weak and weathered rocks, the SPT is affected by many factors and, therefore the interpretation of its results will be uncertain and imprecise. Yet it may be noted that virtually all methods of sampling and testing these types of material are equally difficult. For example, rotary core sampling often proves difficult or impossible. Under such conditions the SPT may be an invaluable tool in reassuring the engineer that the ground actually is as competent as was thought before the start of the ground investigation.

6.4 SYNOPSIS

1. The SPT is a small diameter, dynamic test to failure. The N value will reflect these factors.

2. In granular soil, a great many factors inter-react to produce the measured penetration resistance. Simple correlations, which do not consider more than one factor, are likely to be unreliable.

3. The major factors controlling dynamic penetration resistance in granular soils are:
   - cementing
   - ageing
   - mean effective stress level
   - overconsolidation
   - relative density
   - particle size.

4. There is a clear need for explicit standardisation of density descriptors used on the basis of SPT N value as part of sample description. The system recommended here uses N values corrected for both energy and overburden pressure. Whatever system is adopted, its basis must be clearly stated.

5. In cohesive soils, undrained shear strength is the main factor controlling penetration resistance. The measured value of undrained shear strength obtained from other in-situ or laboratory tests depends upon test size and method, thus affecting correlations with the SPT.

6. Correlations between undrained shear strength and SPT N will depend upon:
   - the method of measuring c_u
   - the sensitivity of the soil
   - fissuring.

   Available evidence suggests that the SPT provides a cheap reliable method of determining the undrained shear strength of firm to hard clays.

7. The factors controlling dynamic penetration resistance in weak and weathered rock are:
   - the strength of the intact rock
   - the porosity of the intact rock
   - the spacing, aperture and tightness of joints
   - the presence of hard inclusions, such as flint or chert.

8. The SPT cannot be used to provide reliable estimates of the visual weathering grade of chalk.

9. Although the SPT cannot be expected to provide precise estimates of weak rock parameters, it is one of very few methods available. Therefore, in order to enhance its reliability, site specific correlation with directly determined parameters should be carried out whenever possible.

7 SPT applications

The Standard Penetration Test has been used for many purposes. At its simplest, it is a low-quality sampler. At its most useful it is a rapid, inexpensive, qualitative test which can provide data even when other techniques of sampling or testing are not viable or cannot be justified financially.

The earliest application, the prediction of the allowable pressure of shallow foundations on sand, fulfilled a major need. Increasing use of the test then led to attempts to correlate its results with other geotechnical characteristics. Starting from site-specific correlations, more general ones have been proposed. But all too often specific correlations have been applied to inappropriate circumstances – in spite of the warnings of the original authors that such correlations were of limited validity.

From the previous chapters it can be seen that the SPT will often yield data which are, to some extent, affected by factors other than soil conditions. It is not, then, a test which yields data of great subtlety and precision. But this is not to deny its usefulness, given reasonable care in the pre-test work and testing according to sound established principles.

1. For many types of ground the SPT gives a basis for broad, practical judgements of properties and likely behaviour. For example, if penetration resistance is high (say greater than 30 to 50), then there are not likely to be foundation difficulties, and there will be no need for more refined geotechnical studies if ordinary types of construction are envisaged. Similarly, more refined geotechnical calculations will not be required if blowcounts are low (e.g. 0 to 4 or 5) since ordinary types of spread foundation cannot be used.

2. When intermediate conditions are encountered and more extensive, refined calculations are required, it is important that both the test and computational methods available are judged in relation to the required accuracy of the prediction. This judgement should be made of all methods, whether SPT-based or not, and should consider the following aspects:

    (a) The probability of there existing a reasonable relationship between a particular test result and the geotechnical parameter controlling the behaviour of the ground.

    (b) The availability of evidence of accuracy from previous applications of a test in similar ground conditions. In this respect it is wise to remember that engineers rarely report their failures (see, for example, van Weele, 1989).

    (c) The levels of sensitivity and sophistication of a given test in relation to the analysis to be performed. It is, for example, as incongruous to use the Self-Boring Pressuremeter to derive undrained strength values of clay for bearing capacity analysis as it would be to use the SPT to obtain shear modulus values for input into an anisotropic non-linear elasto-plastic finite element analysis.

3. If methods of calculation are to be used which derive from site-specific correlation, it is essential to consider whether ground conditions are sufficiently similar to those at the site where the correlation was developed. Can correlations developed in sand sensibly be used in gravel, for example, or can correlations developed in clay be used in chalk? In all probability they cannot.

4. Where methods have been developed outside of the UK, then different techniques of drilling, different equipment and different test techniques will have been used. The relationship between a given variable and the SPT penetration resistance will not be the same when using UK equipment and techniques (Clayton, 1986).

Applications of the SPT in the solution of geotechnical problems may be divided into two groups (Table 7). Direct methods of analysis take the test result (in this case, N) and proceed directly to the required quantity (e.g. settlement or allowable bearing capacity). Indirect methods, on the other hand, use the test result to derive geotechnical parameters (e.g. compressibility or strength) and these may then be used with any suitable form of geotechnical analysis to determine the required quantity. In the following sections these applications are considered individually, and in detail. Wherever possible, it will be wise to use the SPT as part of an indirect method, since this will allow the geotechnical parameters so derived to be critically examined before being used in a design calculation, and will also allow the most appropriate design theories to be used.

Table 7 *Examples of direct and indirect applications of the SPT in geotechnical design*

Indirect applications:	**determination of geotechnical parameters**

- effective angle of friction of sand
- undrained shear strength of clay
- unconfined compressive strength of weak rock
- equivalent Young's modulus of granular soils
- maximum shear modulus, G_{max}

Direct applications:	**design calculations**

- settlement of spread footings on sand
- allowable bearing pressure of footings on sand
- allowable bearing pressure of rafts on sand
- liquefaction potential of sands
- shaft and end resistance of piles
- liquefaction of sands
- sheet pile driveability

But the SPT is also useful for the more fundamental processes of profiling and soil classification, as described in the following sections. These are essential features of any site investigation, during which the spatial variability of the ground is investigated, and initial estimates of its engineering behaviour are made.

For all these applications, a large sample of SPT results is desirable. If there is an inadequate sample of SPT results, or if the end-product is over-sensitive to the input value of N, then SPT-based design methods will not be appropriate.z

7.1 PROFILING

Profiling is carried out during ground investigation in order to establish boundaries between different soil or rock types. A number of tools are commonly used for profiling, both *in situ* and in the laboratory. These include dynamic sounding, cone testing, SPT testing and, in the laboratory, index testing using moisture content and undrained shear strength or unconfined compressive strength tests.

Dynamic penetrometers of different sizes, used to different specifications, are now common in the UK. The object of testing is simply to make crude estimates of penetration resistance at different depths and locations across a site. At shallow depth, profiling is more economically carried out using light-weight penetrometers such as those conforming to German Standard DIN 4094. These penetrometers may not be able to drive past minor obstructions, such as flints in chalk, which will easily be pushed aside or removed during boring or SPT testing. It may well be, then, that part of the ground will be profiled using light-weight penetrometers, while another part is profiled using heavier penetrometers, including the SPT.

Penetration resistance may be converted from one standard to another by considering the energy delivered to the end area of the penetrometer, and the length of drive, or in other words by plotting the results for different penetrometers in terms of energy/displaced volume of soil, *E/V*, i.e.

$$E/V = \frac{N . E_r}{A.l}$$

where N is the number of blows for drive length l

E_r is the average energy delivered per blow (as distinct from the theoretical free-fall energy)
A is the cross-sectional area of the penetrometer shoe
l is the drive length.

For the SPT carried out in the UK, the drive length, l, is 300mm, and E_r = 70% of 48.5 kgf-m = 34 kgf-m. Most dynamic penetrometers use a solid cone, so that the value of the cross-section area to use is obvious. For the SPT the value is not clear, since it might either be the gross area (i.e. including the central hole by which a sample is taken) or the area of steel. But it is common to find that the sample inside the SPT is shorter than 450mm, suggesting that sample jamming (Clayton *et al.*, 1983) is taking place. When this occurs the penetrometer acts as if it has a cone end. Therefore it is suggested that A should be the gross area, 2027 mm².

When used as a dynamic penetrometer, the SPT will need to be carried out frequently, say at 1 metre centres, in each borehole. It could, of course, be driven continuously ahead of the borehole, but this procedure will tend to give a higher penetration resistance as soil adhesion increases on the drive rods.

Table 8   *SPT-based soil and rock classification systems*

Sands	$(N_1)_{60}$ 0-3	Very loose
	3-8	Loose
	8-25	Medium
	25-42	Dense
	42-58	Very dense
Clays	N_{60} 0-4	Very soft
	4-8	Soft
	8-15	Firm
	15-30	Stiff
	30-60	Very stiff
	> 60	Hard
Weak rock (except chalk)	N_{60} 0-80	Very weak
	80-200	Weak
	> 200	Moderately weak to very strong
Chalk	N_{60} 0-25	Very weak
	25-100	Weak
	100-250	Moderately weak
	> 250	Moderately strong to very strong

Note:
N_1 is SPT N value corrected to 100 kPa effective overburden pressure
N_{60} is SPT N value corrected to 60% of theoretical free-fall hammer energy
$(N_1)_{60}$ is SPT N value corrected for both vertical effective stress and input energy

7.2 CLASSIFICATION

The major strength of the SPT is in its ability to classify a wide range of soil types. Classification is the process used during ground investigation to divide soil and rock into a limited number of groups, each of which contains materials expected to have broadly similar engineering behaviour. The engineering parameters which are of most importance in estimating behaviour are strength, compressibility and permeability and rate of consolidation.

The methods most commonly used for classification are sample description, moisture content and plasticity testing (for cohesive soils) and particle size distribution (for granular soils).

Classification with the SPT is made possible because the test combines both a sampler (albeit of very poor quality and unable to sample coarse granular soils) and a penetrometer. Table 8 shows simple classification systems for four different material types, based upon visual description combined with penetration resistance. These can be used as a basis for preliminary design decision, but should not be used thereafter.

The recent introduction of friction measurement at the end of a Standard Penetration Test, termed the SPTF (Ranzini, 1988), has considerably increased the potential of the SPT for soil classification. Upon completion of the 450 mm SPT drive, the split spoon is turned by means of a 80 kg.m torque wrench applied to the drill rods at ground surface, and the maximum torque, T, is determined. Ranzini proposed that through the equation a measure of side friction

$$S = \frac{T}{(40.5366 \cdot h - 17.4060)}$$

where: S is the shear stress on the side of the split spoon (kgf/cm²)
T is the measured torque (cm.kgf)
h is the split-spoon penetration (cm),

could be obtained which might, for example, be useful in evaluating the side friction of piles. Decourt (1991) has adopted a more pragmatic approach, simply comparing the measured torque with the N value. For an input rod energy ratio of about 70% (i.e. similar to that thought to be delivered by British automatic trip hammers) he found that

1. For sandy alluvial soils the torque (in kg.m) is approximately equal to the N-value (T $\cong$ 2 + N_{70} or T $\cong$ 1.1 N_{70}). For a small group of results on gravelly sand (Figure 35a) he found that the penetration resistance was approximately doubled, thus demonstrating the influence of coarse particles on penetration resistance.

2. For residual soils the torque is much greater ($T = 1.7 N_{70}$) (Figure 35b).

This simple extension of the test could be developed to produce site-specific correlations with soil type, in much the same way that the friction cone is used. In addition, friction measurements might be used to detect when coarse particles or cementing are giving unreasonably high penetration resistances, and to examine the variability of rod energy ratio, in a suite of tests in similar soil.

7.3 PARAMETER DETERMINATION

As noted above, in indirect methods of analysis the SPT is used to determine a soil parameter (for example Young's Modulus) which is then used as an input to a suitable method of analysis. This method of working with SPT results is to be preferred because the parameters determined from N values can be examined to see whether they appear reasonable, that is whether they

- fall in the range expected for the soil type

- give results which are broadly in agreement with those obtained from other tests.

The use of the SPT to obtain soil parameters has been increasing over recent years. Although early methods were typically direct (as, for example, in the derivation of allowable bearing pressure for footings, see (Terzaghi and Peck, 1948), the past two decades have seen the derivation of correlations for a wide range of parameters, from the undrained shear strength of clays to the maximum shear modulus of sands.

To the unwary, parameters determined from the SPT may appear to be of some accuracy. While it is true that the SPT can provide a more accurate and reproducible value of the undrained shear strength of an insensitive clay than the undrained triaxial test, most other parameters will be less well determined by the SPT than by other tests, whether laboratory or *in situ*. Therefore, when parameters are required with reasonable accuracy, either for reasons of economy or safety, it will be necessary to check the values using other test methods.

Soil mechanics design calculations often give the impression of being both scientifically based and analytical in nature. Many methods, despite using apparently 'real' parameters, are semi-empirical in nature. The parameters used in the calculations are test and test-method dependent, and the use of a different test method may make the design method unreliable. When substituting SPT-derived parameters, caution is required.

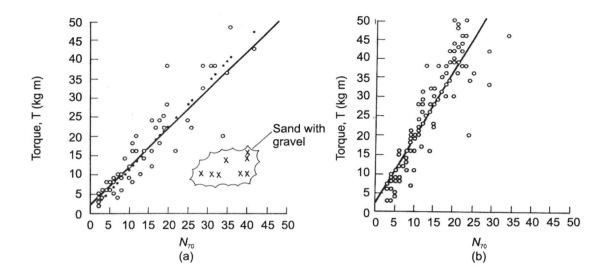

Figure 35 Relationships between torque, T, and SPT value, N_{70} for (a) sandy alluvial soils and (b) residual soils derived from migmatites, granites and gneiss (Decourt, 1991)

Parameter determinations are discussed and explained in Section 8.

7.4   DIRECT DESIGN METHODS

Direct design methods based on the SPT use N values to obtain a final calculated value, such as the settlement of a foundation, without obtaining values of relevant soil parameters as an intermediate stage. Well known examples of this type of calculation are the Terzaghi and Peck methods for estimating allowable bearing pressures for spread foundations and for rafts. More recent examples include Burland and Burbidge's, and Schultze and Sherif's methods for estimating the settlements of spread foundations on granular soils.

Direct design methods are discussed and explained in Section 9.

Direct design methods often have the advantage of simplicity, but they are less easy to check than indirect methods since the soil parameters are implicit. One approach, suggested by Simons and Menzies (1977), is to carry out the same design using a number of methods, and then compare the results. Even this, however, is not particularly satisfactory; having derived several values of settlement (for example) it is impossible to know which one is the most reliable.

8 Determination of geotechnical parameters

The Standard Penetration Test is commonly used to obtain parameters for input into routine geotechnical design calculations. A wide range of parameters, for almost all soil and weak rock types, can be obtained with ease and convenience, and at modest cost.

When using the correlations described below, it is advisable to check the value of each parameter obtained from the SPT with both the expected range of values for the given ground conditions, and with values of the same parameter obtained by other means (for example, from other *in-situ* or laboratory tests). It is also important to examine the factors which control the SPT N value, in order to assess (albeit qualitatively) the likely reliability of the correlation in use.

Table 9 gives a list of parameters which have been correlated with Standard Penetration resistance. A major review of the available correlations was published by Stroud in 1989.

Table 9   *Determination of parameters from SPT results*

Parameter	Material type				Required input
	Granular soil	Cohesive soil	Weak rock	Chalk	
ϕ'	*				$(N_1)_{60}$
c_u		*	*		N_{60}
σ_c			*	*	N_{60}
E_u		*			N_{60}
E'	*	*	*	*	N_{60}
m_v		*			N_{60}
G_{max}	*				$(N_1)_{60}$

Note: N_1 is SPT N value corrected to 100 kPa effective overburden pressure
       N_{60} is SPT N value corrected to 60% of theoretical free-fall hammer energy
       $(N_1)_{60}$ is SPT N value corrected for both vertical effective stress and input energy

8.1 ESTIMATION OF PARAMETERS IN GRANULAR SOILS

8.1.1 Effective angle of friction ϕ'

(a) Suggested method

The two most common applications of the Standard Penetration Test are the determination of the strength and compressibility of sands and gravels. It is normal to determine the effective angle of friction, ϕ', and then to use this value in design calculations. On the other hand, compressibility is often implicitly determined as part of a direct design method (see Section 7.4.1).

The effective angle of friction of both sands and gravels is most frequently determined, in the UK, from Peck, Hanson and Thornburn's chart (Peck et al., 1974, Figure 36(a)). There are good reasons (see Discussion, below) to believe that this chart will provide very significant underestimates of the effective angle of friction, and therefore its continued use can be recommended as a basis for routine design. The measured penetration resistance should be corrected both for energy and overburden pressure before obtaining the effective angle of friction from Figure 36(a). A further correlation, which explicitly recognises the influence of vertical effective stress, has been proposed by Mitchell et al. (1978), and is shown in Figure 36(b).

(b) Discussion

It is known that the effective angle of friction of granular soil is a function of a considerable number of factors, amongst which density, grain characteristics (e.g. particle size distribution, particle angularity) and applied effective stress level are considered very important. Dynamic penetration resistance is also significantly affected by these factors, but the inter-relationships are not known in detail.

The value of effective angle of friction required for a particular design application may possibly be the triaxial (i.e. axi-symmetric) value, but for most engineering problems (for example retaining walls, slopes, and long footings) it is clear that it is the plane strain value that is required. The plane strain angle of friction normally exceeds the triaxial value by a considerable margin (see, for example, Cornforth, 1973; Bolton, 1986). Although Cornforth's work suggested that the two values are the same at zero relative density, in most practical situations the plane strain angle of friction may lie some 10% above the triaxial value (for example, see Wroth, 1984 and Bica and Clayton, 1989). Most commonly, values of effective angle of friction have, in the past, been determined in the triaxial apparatus.

The strength of granular soil may be derived from at least four principal sources; the critical state angle of friction, ϕ_{cv}' (the value obtained at approximately zero relative density), a dilatancy component, $(\phi' - \phi_{cv})$ which varies with degree of interlocking (and therefore relative density), the effective normal stress, and cementing. Figure 37, from Bolton (1986), shows how relative density and the dilatancy component of effective angle of friction are related for uncemented and unaged quartz sands. For these sands it was found that the critical state angle of friction, ϕ_{cv} varied from about 33° for quartz sands to 37° for feldspathic sands. Youd (1973) found that ϕ_{cv} varied with particle roundness, from about 30° to 36° for uniformly graded quartz sands, from 32-39° for sands containing feldspar, and from 35-41° for well-graded sands.

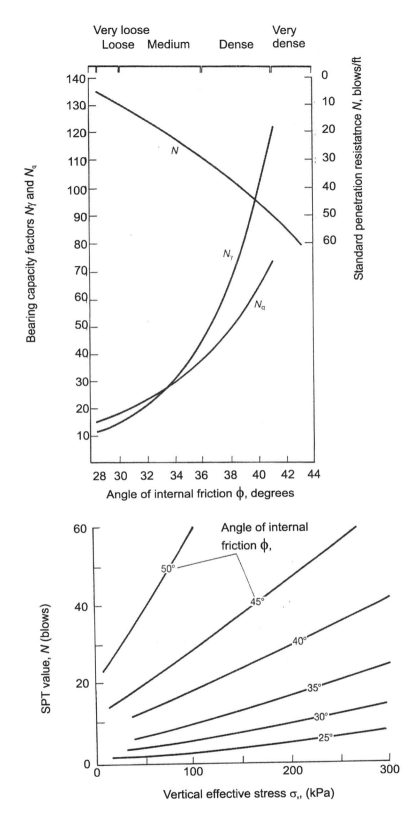

Figure 36 *Suggested relationships between (a) (N, φ' and bearing capacity factors $N_γ$ and N_q (Peck et al., 1974) and (b) N, φ' and $σ_v'$ (Mitchell et al., 1978)*

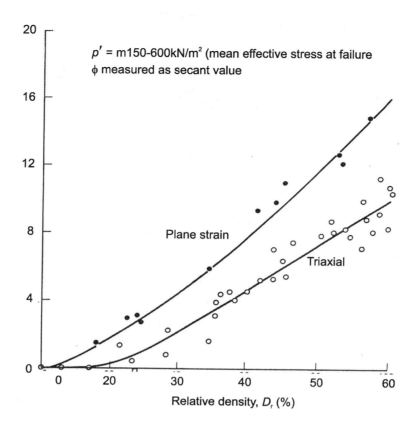

Figure 37 *Influence of relative density on the dilatancy component of effective angle of friction (Bolton, 1986; Stroud, 1989)*

As a result of the geological processes involved in particle degradation, coarser soils are typically more well-graded and have greater particle angularity. Data are not numerous because of the problems of sample preparation and testing, but 300mm shear box tests certainly suggest direct shear angles of friction for sands and gravels which vary from 40° to 60°.

It can therefore be concluded that the plane strain angle of friction of the granular soils typically encountered in the field may vary from about 35° to 60°. Peck et al., chart gives values ranging from 28° ($(N_1)_{60}$ = 3) to 43° ($(N_1)_{60}$ = 60). Even given the lack of a relationship between N and ϕ, particularly at low relative densities (Stroud, 1989), for sands, the confusing inter-relationships of stress level, overconsolidation ratio, and cementing with penetration resistance and strength, Peck et al., relationships can be expected to be conservative.

8.1.2 Stiffness

(a) Suggested method

Although the settlement of small foundations or rafts relatively insensitive to differential settlement has often been estimated using direct design methods, the evaluation of modulus before the prediction of deformations or settlements is now more common.

Many correlations have been published which claim a relationship between stiffness and SPT N value for granular soils. Figure 38 shows a range of relationships given by previous authors, and it can clearly be seen that no consistent pattern emerges. However, it is now accepted that it is necessary to use local strain measurement if laboratory determinations of stiffness are to be reliable and, in addition, the use of local strain testing has consistently shown that in triaxial tests to failure stiffness is a function of strain level, at least down to 10-5 strain. It can therefore be accepted that most previous laboratory determinations of stiffness are likely to be unreliable, yielding stiffness values which are too low.

Stroud (1989) adopted a more logical approach in bringing together the data on the stiffness of granular soils. Using available field data (Burbidge, 1982) giving the settlement of spread footings, raft foundations and plate tests, he recognized the importance of strain level by plotting the ratio E'/N_{60} as a function of 'degree of loading' q/q_{ult} (Figure 39). Here the penetration resistance is corrected to 60% rod energy ratio, but is not corrected for stress level since it is argued that both stiffness and penetration resistance are increased by an increase in mean normal effective stress. Values of q_{ult}, on the other hand, were calculated using N values corrected for both energy and stress level (see Appendix B of Stroud, 1989).

Figure 39 shows that, for both normally consolidated and overconsolidated soils, with a factor of safety of 3 on bearing capacity (i.e. $q_{net}/q_{ult} = 1/3$) a reasonable approximation is

$$\frac{E'}{N_{60}} = 1 \text{ (MPa)}$$

But, judging from Stroud's re-analysis of the available case records, most foundations have a factor of safety considerably in excess of 3, so that for normally consolidated sands E'/N_{60} may rise to about 2 MPa and for overconsolidated sands and gravels it may rise to 16, at very small strain levels.

Burland and Burbidge (1985) have shown that the time-dependent settlements which occur following the completion of loading are considerable, and depend not only upon time but also upon the style of loading (i.e. static, cyclic or dynamic). Since most case records relate to a relatively short duration of loading, when compared with the expected life-span of a typical structure, the immediate settlements calculated using $E'/N_{60} = 1$ should be increased in accordance with the recommendations given in Section 9.1

While for many routine foundation displacement problems it is necessary to determine stiffness at some intermediate strain level, for problems involving soil dynamics, e.g. soil behaviour during earthquakes and beneath vibrating machinery, it is the maximum (very small strain) value of shear modulus (G_{max}) that is required. As noted by Schmertmann (1978) SPT sampler penetration involves primarily dynamic soil shear behaviour, albeit at the failure reference level of shear strain, and therefore it may be reasonable to expect a correlation between N values and G_{max}. Many correlations now appear in the literature (for example, Imai and Yokota, 1982, Shioi *et al.*, 1981, Muromachi *et al.*, 1974, Ohta and Goto, 1978, Imai and Tanouchi, 1982, Sykora and Stokoe, 1983). A comparison of these correlations has been made by Crespellani and Vannucchi (1991), and the relationships are shown in Table 10.

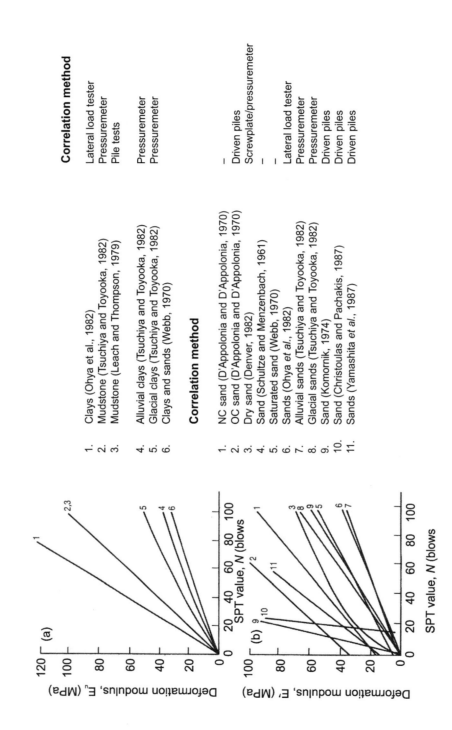

Figure 38 *Examples of correlations between equivalent drained and undrained modulus values and penetration resistance, N, for mudrocks, clays and granular soils*

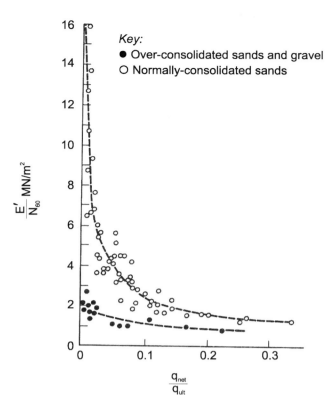

Figure 39 *Relationship between stiffness, penetration resistance and degree of loading for sand (after Stroud, 1989)*

Table 10 *Comparison of some correlations between G_{max} and N (from Crespellani and Vannucchi, 1991)*

	Soil type	a	b	r
Clay	(alluvial)	10.4	1.070	0.500
	(alluvial)	17.3	0.607	0.715
	(glacial)	24.6	0.555	0.712
	(alluvial)	16.0	0.710	0.921
Sand	(alluvial)	12.3	0.611	0.671
	(glacial)	17.4	0.631	0.728
Gravel	(alluvial)	8.1	0.777	0.798
	(glacial)	31.3	0.526	0.552

$G_{max} = \alpha N^b$ (MPa)   r = correlation coefficient

(b) Discussion

A considerable amount of manipulation of the data was necessary in the production of Figure 39. Burbidge's data base is an international one (see, for example, Clayton, 1986) and therefore differing drilling disturbance and rod energy effects must be expected to have had a significant effect on N values. In processing the data, Stroud took some account of what he believed to be the rod energy ratios likely at the time of testing. But in addition, in order to apply his methodology, he required a value of ultimate bearing capacity, q_{ult}, which in itself is subject to considerable uncertainty.

However, the values obtained are similar to those that can be derived from Burland and Burbidge's I_c values (Table 11) which are for immediate settlement. The dependency of E'/N seen in Table 11 is probably removed from Stroud's results (Figure 39) as a result of plotting against 'degree of loading'.

Table 11 Young's modulus derived from Burland and Burbidge's I_c values

Penetration resistance (blows/300mm)	E'/N (MPa) at		
	mean	lower limit	upper limit
4	1.6 – 2.4	0.4 – 0.6	3.5 – 5.3
10	2.2 – 3.4	0.7 – 1.1	4.6 – 7.0
30	3.7 – 5.6	1.5 – 2.2	6.6 – 10.0
60	4.6 – 7.0	2.3 – 3.5	8.9 – 13.5

8.2 ESTIMATION OF PARAMETERS IN COHESIVE SOILS

8.2.1 Undrained shear strength

(a) Suggested method

For insensitive fissured overconsolidated clays, the undrained shear strength equivalent to that from a 100mm diameter specimen can be obtained using Stroud's correlation, i.e.

$$c_u = f_1 N_{60}$$

Values of f_1 depend slightly upon the plasticity of the clay (Figure 31). The penetration resistance should be corrected for rod energy, but not for stress level.

Undrained strengths obtained in this way will give good estimates of the mean undrained strength v. depth profile, taking into account fissuring. They are equivalent to values determined from 100mm diameter specimens. If, as sometimes occurs, the deposit is not fissured then the resulting strength profile will be an under-estimate of the actual value, perhaps by a factor of 1.5 to 2. If the clay is sensitive, this method will also yield an underestimate of undrained shear strength.

The use of undrained shear strength in geotechnical design requires some care, since the value needed varies from application to application. For example, for estimating the earth pressure against braced excavations, or the shaft adhesion on piles, the undrained shear strength equivalent to that obtained from a small diameter (38mm) test specimen is needed. For bearing capacity or short-term (undrained) first-time slope failure the shear strength equivalent to a larger diameter specimen, containing sufficient fissures to be representative of the mass, is required. As Marsland (1972) has shown, in fissured soils these two strengths can be markedly different (Figure 40). The undrained strength derived using Stroud's method will approximately correspond to the mass strength.

(b) Discussion

The determination of undrained shear strength on the basis of SPT N value would seem to be one of the most reliable uses of the test. Stroud's data base contains a very large number of results from boreholes in a variety of cohesive soils, and both the penetration resistances and the undrained shear strengths in the data base were obtained in a reliable and consistent manner.

8.2.2 Compressibility

(a) Suggested methods

As in the case of granular soils, the compressibility and stiffness of cohesive soil is strongly strain level dependent. But in addition, the compressibility of the soil is influenced by the relative rates of loading and drainage of excess pore pressures.

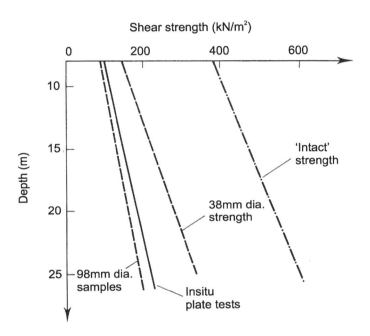

Figure 40 *Comparison of the mean undrained shear strength-depth profiles in the London Clay determined from different test sizes (after Marsland, 1972)*

The compressibility or stiffness of cohesive soil is commonly expressed in a number of ways:

- compression index (C_c)
- coefficient of volume compressibility (M_v)
- undrained Young's modulus (E_u)
- drained Young's modulus (E').

The compression index is routinely used in the calculation of settlements of normally and lightly overconsolidated clays. The predicted compression of such materials is strongly dependent on the value of pre-consolidation pressure used in the calculations, and this cannot be determined from the SPT N value. Furthermore, since the undrained shear strength of such soils will be low, the penetration resistance will not exceed about 10 blows/300mm, and the N value will be an insensitive index of compressibility. Therefore the use of the SPT to predict settlements of construction on normally or lightly over-consolidated clays is not recommended.

In the UK, the most common technique for the settlement of overconsolidated soil is Skempton and Bjerrum's (1957) method. Values of undrained Young's modulus, E_u, and coefficient of volume compressibility are required.

The coefficient of volume compressibility of stiff fissured clay has been correlated with the SPT N value by Stroud and Butler (1975). The value can be obtained from the equation

$$m_v = \frac{1}{f_2 N}$$

where f_2 is obtained from Figure 41. The correlation given in Figure 41 was obtained from comparisons of 76mm diameter 19mm high oedometer test results with SPT N values, for British soils. As a result of improved laboratory instrumentation, it has been possible to demonstrate the very large effect that bedding has in reducing the measured compressibility of overconsolidated clays. There must now be considerable doubts as to whether the specimen configuration used in the oedometer can produce realistic parameters for design. There is consequently a trend towards the use of undrained and drained modulus values in design, not only for the prediction of retaining wall movements, but also for the prediction of settlement and heave.

Most heavily overconsolidated clays are significantly anisotropic, principally because of differences between the vertical and horizontal *in-situ* effective stresses they sustain. Typically, near to ground surface, horizontal effective stress exceeds the vertical, and the horizontal stiffness will be greater than the vertical. The influence of stiffness anisotropy can be taken into account by deriving average values of isotropic moduli from field data, but the different directions of loading and unloading may have to be reconsidered when selecting a realistic value for design.

Based upon case histories of foundation, Butler (1975) showed that good estimates of the undrained settlement and heave of structures on stiff fissured clay could be obtained by using an isotropic undrained stiffness in elastic calculations of settlement which was equal to 220 times the undrained shear strength determined from a 100mm diameter triaxial specimen, together with Poisson's ratio of 0.1. If the undrained shear strength is obtained from the SPT N value then, from Figure 31 depending upon the plasticity of the clay.

$$\frac{E_u}{N} = 1.0 - 1.2 \text{ (MPa)}$$

Data from Stroud shows, again using case histories, that this relationship is likely to be reasonable for design for a wide range of 'degree of loading' (q/q_{ult}), at least down to about 0.1. Below this, stiffnesses rise towards the very small strain values for which

$$\frac{E_u}{N} = 6.3 - 10.4 \text{ (MPa)}$$

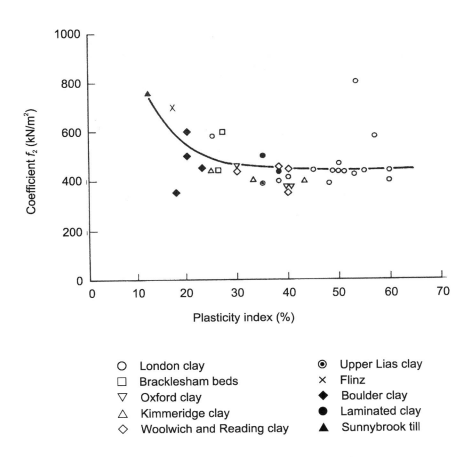

Figure 41 *Correlation between coefficient $f_2 = \dfrac{1}{m_v N}$ and plasticity index (after Stroud and Butler, 1975)*

When estimates of horizontal undrained movements are to be calculated, higher values of undrained Young's modulus have to be used since the stiffer horizontal Young's modulus will tend to dominate behaviour. For the particular case of the London Clay, for example, the horizontal Young's modulus is between 1.5 and 2.0 times greater than its vertical value.

The relevant drained modulus for use in design is also a function of strain level and anisotropy. Since
the undrained and drained Young's modulus of an isotropic material can be related Poisson's Ratio
($E'/E_u = (1 + v)/(1 + v_u)$) it is expected that, for a drained Poisson's ratio of 0.1, $E = 0.73\, E_u$.

Butler (1975) showed that, predominantly for structures founded on the London Clay, $E' = 130\, c_{u(100)}$. Therefore

$$\frac{E'}{N} = 0.6 - 0.7 \ (\text{MPa})$$

Stroud (1989) reworked data from a number of case records, on a range of clays, and showed that a somewhat higher value

$$\frac{E'}{N} = 0.9 \text{ (MPa)}$$

was appropriate from $q/q_{ult} = 0.4$ down to values of q/q_{ult} of the order of 0.1, with higher ratios being observed below this (Figure 42). Once again, the modulus should be increased when considering horizontal loading or unloading.

(b) Discussion

The factors controlling the stiffness of cohesive soils are now well understood. Although more satisfactory correlations are obtained between mean effective stress and soil stiffness, good correlations between stiffness and undrained strength can be obtained for soils at similar strain levels and following the same stress path.

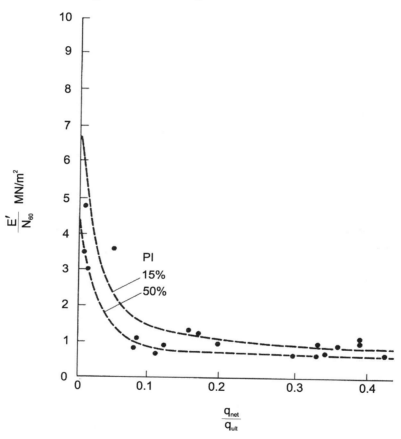

Figure 42 *Relationship between effective modulus of overconsolidated clay, penetration resistance and degree of loading (after Stroud, 1989)*

Therefore, since the SPT provides a reliable measure of the undrained shear strength of insensitive overconsolidated clays, and since the proposed relationships are based upon field observations of full-scale structures, there is good reason to have confidence in their ability to provide conservative design parameters.

These relationships should, however, be used only for simple, unsophisticated, design work. For complex soil-structure interaction problems it is necessary to obtain parameters in a number of ways, but principally from high quality triaxial tests instrumented with local strain and mid-plane pore pressure measuring devices. *In-situ* pressuremeter tests will also usually be required to provide values of horizontal total stress, and these can be used to provide values of stiffness in the horizontal loading direction.

8.3 ESTIMATION OF PARAMETERS IN WEAK ROCK

8.3.1 Strength

(a) Suggested method

The SPT is used in weak rocks both to obtain estimates of uniaxial unconfined compressive strength (σ_c), and for obtaining operating strengths that take into account rock mass characteristics, such as jointing, for direct input into design. At shallow depth most rocks are fractured, and the operational strength required for design will then depend upon the scale of the problem relative to the jointing of the rock mass, and the way in which shear stress is to be applied; but the operational strength will generally be less than the shear strength obtained from unconfined compressive strength tests of the intact rock. As depth increases, the importance of fracturing decreases, and the strength of the intact rock becomes more significant.

Based upon pile and pressuremeter test results, Figure 43, it has been found that the relationship for clays, i.e.

$$c_u = 5 \cdot N_{60} \text{ (kPa)}$$

also continues to be approximately valid for weak rocks, although it appears rather conservative for materials with an unconfined compressive strength ($\sigma_c = 2c_u$) greater than about 4 MPa. The unconfined compressive strength will therefore be

$$\sigma_c > 10 \cdot N_{60} \text{ (kPa)}$$

The application of these parameters is often to the design of permissible end-bearing pressure of driven piles on competent rock, or to the design of shaft and end-bearing resistance of piles in weak rock.

(b) Discussion

The strength of a weak rock estimated from SPT N values cannot be expected to be of great accuracy. In rocks the SPT will typically be terminated after 50-100 blows, so that the value used in design must be extrapolated. And there is evidence (Meigh, 1980) that the rock material type has an influence on penetration resistance (as can be seen by comparing the correlation given here with that for chalk, Section 8.4.1). Figure 43 is plotted to a double-logarithmic scale, and it may be observed that even for N_{60} values less than 200 blows/300mm there is a scatter of about 20% about the mean. But for higher values of unconfined compressive strength even greater divergence has been observed. Finally, it should be recognised that the solid cone may give higher penetration resistances than the open shoe. Since early correlations (e.g. Cole and Stroud, 1976) were based upon N values determined with the open shoe, it is this tool which is recommended for routine use.

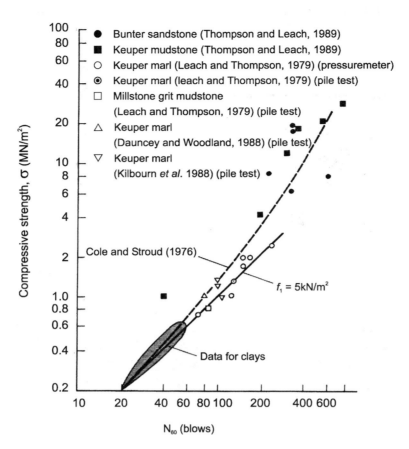

Figure 43 *Correlations between unconfined compressive strength and penetration resistance of weak rocks (after Stroud, 1989)*

Despite the fact that the proposed method is probably limited in accuracy, it can be recommended on the basis that it has been proven, as far as practical, by its application to real sites in a range of weak rock materials such as the Mercia Mudstone, the Bunter Sandstone, and the Millstone Grit Mudstone (Leach and Thompson, 1979; Thompson and Leach, 1989; Kilbourn et al., 1989).

8.3.2 Stiffness

(a) Method

The stiffness of weak rocks may be obtained by means of a number of tests, of which the SPT is one. Others include plate testing, Hughes high pressure dilatometer testing, self-boring weak-rock pressuremeter testing and cross-hole or down-hole geophysical testing, as well as laboratory stress path testing with local strain measurement. Each method will yield a different range of results in the same weak-rock deposit, partly because of the direction of loading and the volume of loaded rock, and partly because of the strain levels imposed.

For the estimation of the load/settlement behaviour of piles, Leach and Thompson (1979) observed that $E'/N_{60} = 0.9 - 1.2$ (MPa), a somewhat lower value than had previously been suggested by Stroud (1974). Further field evidence, collected by Stroud (1989), shows that for a wide range of weak rocks (except chalk)

$$\frac{E'}{N_{60}} = 0.5 - 2.0 \text{ (MPa)}$$

and that for a factor of safety against bearing capacity failure of more than 3, E'/N_{60} is likely to exceed 1.

(b) Discussion

Because of the very small strains imposed on the rock mass by shear wave geophysical testing, it is to be expected that this method will yield the highest modulus values. Thompson and Leach (1989) and Thompson et al. (1990) have shown that high pressure dilatometers and rock self-boring pressuremeters yield stiffnesses for weak rocks which are typically between one-half and three-quarters of those given by geophysical methods. In contrast, the ratio between G_{max} (the stiffness determined using geophysics) and the value obtained using empirical relationships based upon the SPT appears to decrease with increasing strength, from about 0.2 at low strength to as little as 0.03 at higher strengths (for example, Thompson et al. (1990)). The reason for this trend is not clear.

Many of the case records used by Stroud are for piles or footings on Keuper Marl. The extent to which the proposed correlation can be relied upon in other deposits remains unknown.

8.4 ESTIMATION OF PARAMETERS IN CHALK

8.4.1 Strength

(a) Suggested methods

The effective angle of friction of the chalk remains more or less constant at about 34° throughout the deposit, and need not, therefore, be the subject of testing. Chalk has, however, a very variable porosity and, as a result, a variable intact effective cohesion intercept and a variable unconfined compressive strength. No data are known which relate the effective cohesion intercept of intact chalk to its penetration resistance; in any case, the influence of jointing means that intact cohesion intercept is rarely of interest in design involving near surface, more weathered chalks.

Hobbs (1977), Hobbs and Healy (1979) and Meigh (1980) give data which can be interpreted in terms of mass strength, and which is largely derived from pile and plate load tests. Hobbs concluded that the end-bearing resistance of piles in chalk could be related to penetration resistance by $q_u = 0.25\ N$ (MPa) for $N < 30$, and $q_u = 0.2\ N$ (MPa) for $N > 40$. Using Hobbs and

Healy's data, Meigh (1980) proposed that the operational undrained strength was approximately 35 N (kPa), or that

$$\sigma_c = 70 \, N_{60} \text{ (kPa)}$$

more recently, but largely based upon the same data, Stroud (1989) has suggested that

$$\sigma_c = 50 \, N60 \text{ (kPa)}$$

these values may be multiplied by a suitable factor (typically 4.5) to obtain an estimate of the ultimate end-bearing resistance of piles in chalk. They should not, however, be used to predict shaft adhesion. For bored piles, Lord (1990) has shown that from case records available capacity is relatively unaffected by the compressive strength and fracturing of the chalk. Shaft adhesion is best calculated on the basis that the horizontal total stress is produced by the column of fluid concrete, that the effective cohesion intercept is zero, and the effective angle of friction is 34°. On this basis, the unit shaft resistance is

$$\tau_{sf} = 0.67 \, (24 - u).z$$

where z is the pile length down to the point at which shaft resistance is to be calculated, and u is the pore pressure in the ground at that depth. The shaft resistance of driven piles will generally be very much lower (see Lord, 1990), and again appears unrelated to chalk properties.

(b) Discussion

The unconfined compressive strength of the chalk is strongly related to the porosity of the material since, with the exception of the Cenomanian, the deposit is chemically homogeneous and has a high purity of calcium carbonate. However, it has repeatedly been shown that penetration resistance in the chalk is a function of drilling technique (Clayton, 1990a; Matthews *et al.*, 1990; Montague, 1990; Twine and Grose, 1990), and that for weathered chalks the penetration resistance is related in a complex way not only to strength but also to fracture spacing, openness, and flint content. Therefore the relationships given above will be unreliable in near-surface chalks, but may prove of greater use in deeper situations, where fractures are infrequent and closed, for example in pile design.

The unconfined compressive strength of intact chalk varies from about 0.5 to 30 MPa. In very high porosity chalk it appears that penetration resistance is unaffected by fracturing, and therefore does not increase with depth. N values of about 6 have been measured, when the unconfined compressive strength of the intact material was about 1.25 MPa. In stronger chalk at depth, Woodland *et al.* (1989) found σ_c to be about 14.5 MPa, while the average SPT N value was determined as 225 blows/300mm.

8.4.2 Compressibility

(a) Suggested methods

Weathered chalk exhibits yielding behaviour (see, for example, Clayton, 1990a) with the normal range of engineering stress. Since, beyond yield, compressibility is typically increased by about a factor of 10, it is important to know the stress level at which yield can be expected to occur. For Grade III, IV, and V chalk, yield typically occurs at between 150 and 500 kPa average applied foundation stress, while for Grades I and II it will not be expected to occur at a stress of less than 1000 kPa. The SPT cannot be used to estimate the value of the yield stress.

Pre- and post-yield moduli are best obtained by large diameter plate testing. Geophysical tests can give a good estimate of pre-yield modulus. But since such methods require special skills, the SPT is frequently used to predict settlements. Wakeling (1970) noted the strain dependency of his back-analysed results, and therefore proposed two relationships, as shown in Figure 44. Many other relationships have been proposed over the years. Reviewing the literature, Stroud (1989) has suggested that conservative (i.e. over-) predictions of the immediate settlement of spread foundations in Chalk can be made on the basis that

$$\frac{E'}{N_{60}} = 5 \text{ (MPa)}$$

this relationship is compared with Wakeling's on Figure 44.

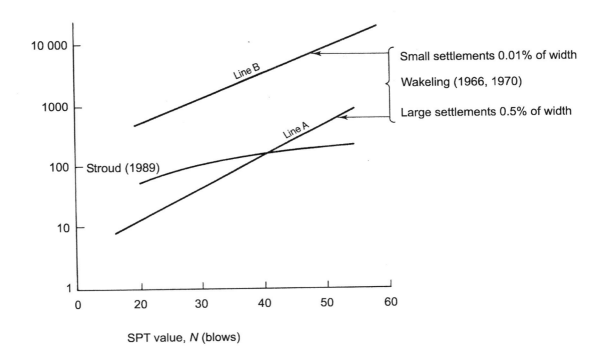

Figure 44 *Correlation of E' with SPT N value for chalk (after Wakeling, 1966, 1970; and Stroud, 1989)*

The data certainly suggest that much higher values are possible, with some larger width foundations yielding E'/N_{60} values up to 45 MPa. It is suggested that no allowance for long-term creep need be made when using these methods.

In reality, the mass modulus of chalk is a function of the modulus of the intact material and the degree to which weathering has then reduced that modulus. The modulus of intact chalk is a function of its porosity and can therefore be related to penetration resistance, provided that the joints in the chalk are widely spaced and tight, which is the case for Grades I and II. For these grades the mass modulus is similar in magnitude to the intact modulus (Burland and Lord, 1970). Although only limited data exist, it would appear that the intact modulus may be estimated for relatively unweathered chalks from

$$E'_i = 50 \, N_{60} \, (\text{MPa})$$

observed values of the stiffness of laboratory specimens of chalk give values of E'/N_{60} ranging from 30-60 MPa for extrapolated N values in low porosity chalks (Woodland *et al.*, 1989), to 160-250 MPa for the Mundford (medium porosity) chalk (Longworth, 1978; Burland and Lord, 1970; and Wakeling, 1970), and 80-190 MPa for very high porosity Suffolk chalk.

(b) Discussion

There can now be little doubt that the SPT N value is a very poor basis upon which to calculate foundation settlements. As has been noted, above, N is strongly affected by drilling methods in the chalk. In addition as weathering becomes more pronounced the ratio between intact and mass compressibility increases, especially in the harder chalks. Since penetration resistance is not related to fracturing in the same way as is compressibility, and continues to some extent to be related to intact strength, its application is unreliable.

Clayton (1990a) showed that, for a particular site, the quoted methods over-predicted immediate settlements by a factor of 7-10 times. The effects of long-term creep would be expected to reduce this over-prediction ratio to the order of 3 – 5. Geophysical methods, and particularly the continuous surface-wave technique (Clayton, 1990a) can be expected to provide better accuracy in chalk.

9 Direct design methods

Design methods of this type proceed directly from the SPT N value, as an input parameter, to the value to be calculated (for example, the settlement of a foundation), without providing any estimate of the soil properties. The earliest use of the SPT in design was for this type of calculation (see Terzaghi and Peck, 1948, for example), although at that stage in the development of the test it was recognised that the values obtained were estimates. These types of design are potentially unhelpful, in that they do not allow the implied soil parameters to be checked, either on the basis of experience or with values obtained in other ways.

In support of direct methods of design, it must be acknowledged that they are often very easy to use (see, for example, Terzaghi and Peck's method for estimating the allowable bearing pressure for spread footings on sand, Terzaghi and Peck, 1948, 1967). Also direct correlation between the desired end product and the input parameters allows the use of statistical methods (for example, multiple regression analysis) to assess the relative importance of the parameters. While design methods based on parameter determination may sometimes appear to provide solutions of a greater accuracy than may reasonably be expected, direct methods will be recognised by most engineers as providing only *estimates* of performance.

9.1 ESTIMATION OF SETTLEMENTS OF SHALLOW FOUNDATIONS ON GRANULAR SOIL

9.1.1 Suggested methods

Bearing capacity will not normally be the most critical factor in the design of shallow foundations, except in the case of very narrow foundations (< 1m wide). Therefore for most design cases it is the allowable bearing pressure, determined from the maximum acceptable settlement, which governs design. But in all cases the safe bearing pressure of the foundation should also, of course, be checked.

The methods of settlement analysis which are recommended in this section are statistically based, drawing on the observed settlements of actual foundations or plate tests. These methods give better accuracy of prediction than more traditional, essentially intuitive, methods. N values are uncorrected, unless otherwise stated.

(a) Initial estimate of foundation settlement

It is often useful to be able to make a rapid, albeit crude, first estimate of the allowable bearing pressure, q_a, or settlement, ρ, of a foundation. Based upon Terzaghi and Peck (1948), a method generally found to be conservative,

$q_a = 10 N$ where q_a is in kPa

and

$\rho = 2q/N$ where q is in kPa

ρ is in mm.

A more informed estimate of settlement can be obtained using a method suggested by Burland et al (1977). The mean SPT value, $\tilde{N}$, is determined by averaging the N-values determined during site investigation for a depth equal to $1.5 B$ below the proposed foundation level, where B is the estimated foundation breadth. Figure 45 shows settlement records plotted as settlement (mm)/applied foundation pressure (kPa) as a function of foundation breadth. An estimate of the likely upper limit of settlement, based upon this plot, is as follows:

Loose ($N<10$) $\rho_{max} = q (0.32 B^{0.3})$

Medium dense ($10 < N < 30$) $\rho_{max} = q (0.07 . B^{0.3})$

Dense ($N>30$) $\rho_{max} = q (0.035 . B^{0.3})$

The probable settlement will be about one-half of max.

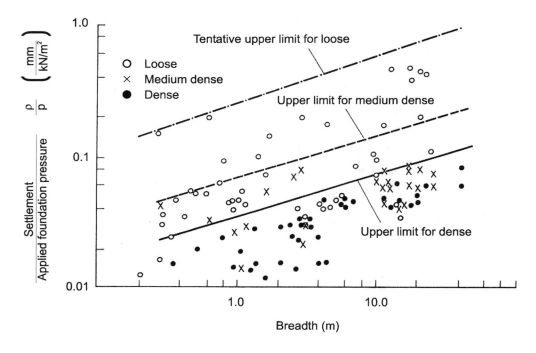

Figure 45 *Observed settlements of footing on sand of various densities (after Burland et al. 1977)*

(b) Detailed analyses

More detailed methods of analysis have been proposed by Schultze and Sherif (1973) and Burland and Burbidge (1985).

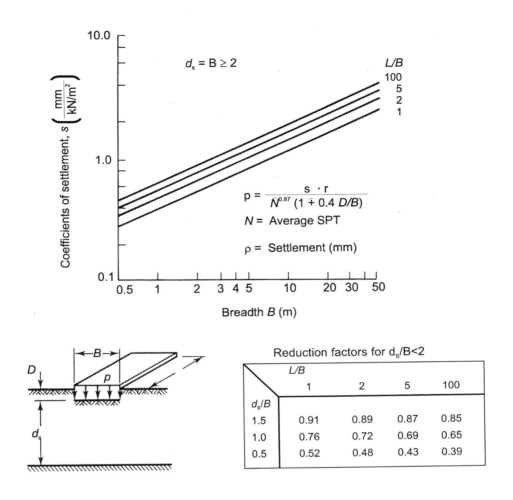

Figure 46 *Schultze and Sherif's method (1973) for calculating the settlement of spread foundations on sand*

Schultze and Sherif's method is based upon multi-correlation statistical techniques, and elastic analysis. For Poisson's Ratio equal to zero, the equation for the settlement of a uniformly loaded area as the surface of a semi-infinite, homogeneous, isotropic elastic mass reduces to

$$\rho = \frac{qBf}{E}$$

where E is Young's Modulus
and f is an influence factor dependent upon the geometry of the foundation.

Using statistical analysis, Shultze and Sherif produced an equation for Young's Modulus based upon the above equation and case records of observed foundation settlements, and from this derived an equation for settlement in terms of N.

The method is summarised in graphical form in Figure 46, where the settlement coefficient, s, is plotted as a function of the breadth of foundation, B, for various length/breadth (L/B) ratios. The values of the influence factor for shallow depths of compressible material are also given in tabular form, and for layered cases the principle of superposition may be used.

Burland and Burbidge's method uses a data base of 100 case records of settlement gathered by Burbidge (1982). Regression analysis was used to derive a relationship between settlement and bearing pressure, breadth of loaded area and average N over the depth of influence. A number of other influencing factors were then examined, using engineering judgement the method detailed below was finally derived.

For a normally consolidated sand the average settlement, , at the end of construction for a surface foundation is

$$\rho = q' . B^{0.7} . I_c$$

where ρ = settlement (mm)
 q' = average effective foundation pressure (q-u)
 B = breadth of loaded area
 I_c = compression index

 = $1.71/N^{1.4}$ (see Figure 47).

In deriving the compression index, I_c, two cases may require the use of corrected N values:

(i) for very fine, silty sand, below the water table,
 $N = 15 + 0.5 (N_{measured} - 15)$

(ii) for gravel or sandy gravel
 $N = 1.25.N_{measured}$.

The mean penetration resistance, $\tilde{N}$, is assessed over a depth of influence, Z_1 below the bottom of the foundation. Figure 47 allows Z_1 to be determined from the breadth, B, when N is constant, or increases with depth. Where N decreases with depth, Z_1 should be taken as $2B$, or the depth to the bottom of the 'compressible' layer, whichever is the lesser.

Overconsolidation or preloading is considered to have a major effect on settlement. If σ_{vo}' is the maximum vertical effective stress at foundation level

$$\rho = (q' - 2/3 \, \sigma_{vo}') \, B^{0.7}.I_c$$

which is equivalent to a threefold reduction in compressibility for stress increases below σ_{vo}'. Thus if σ_{vo}' is not exceeded

$$\rho = 1/3 \, q'. B^{0.7} . I_c$$

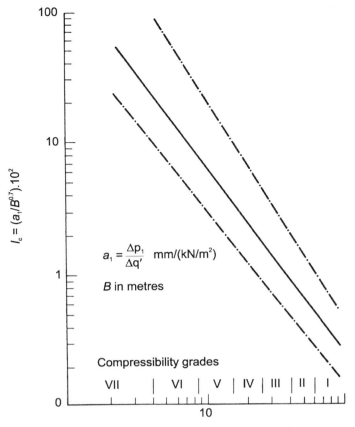

Relationship between compressibility (I_c) and mean SPT blow count ($\bar{N}$) over depth of influence. Chain dotted lines show upper and lower limits

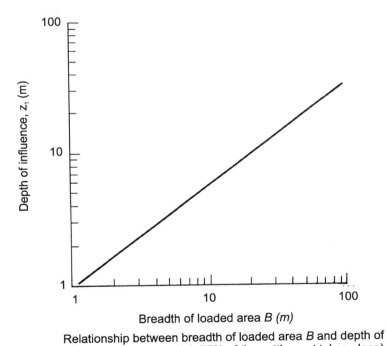

Relationship between breadth of loaded area B and depth of influence z_1 (within which 75% of the settlement takes place)

Figure 47 *Burland and Burbidge's (1985) method for estimating the settlement of granular soils*

Burland and Burbidge concluded that neither depth of founding nor the position of the water table can be shown statistically to influence foundation settlement, but they did infer significant influences due to L/B ratio, thickness of the compressible layer below the foundation, and time. The following influence factors should be used,

(i) for L/B > 1

$$f_s = \left(\frac{1.25 L/B}{(L/B) + 0.25}\right)^2 \text{ for which } f_s \to 1.56 \text{ as } L/B \to \infty$$

(ii) for limited thickness of compressible material (H_s) below the foundation

$$f_l = \frac{H_s}{Z_1}\left(2 - \frac{H_s}{Z_1}\right)$$

(iii) for estimating settlements more than three years after the end of construction

$$f_t = \left(1 + R_3 + R_t \log \frac{t}{3}\right)$$

Where R_3 is the proportion of end-of-construction time-dependent settlement occurring in the three-year period following construction, and R_t is the proportion of end-of-construction settlement taking place during each log cycle of time after three years.

	R_3	R_t
Static loading	0.3	0.2
Fluctuating loading	0.7	0.8

9.1.2 Discussion

Although the methods described above have been shown to be considerable improvements on the twenty or so other methods previously derived for use with the SPT, it should be recognised that their accuracy remains poor. This results from a combination of factors, including the fact that penetration resistance in granular soil is not as greatly influenced by compressibility as by a number of other factors such as stress level, test equipment and method, and drilling method. A large body of literature now exists to demonstrate this (Bratchell *et al.*, 1975; Simons and Menzies, 1977; Talbot, 1981; Milititsky *et al.*, 1982; Jayapalan and Boehm, 1986; Clayton *et al.*, 1988).

For both Burland and Burbidge, and Schultze and Sherif's methods the range of observed/predicted settlements falls roughly between 0.2 and 10 (i.e. between an underprediction of 5 and an overprediction of 10 times the actual observed settlement), although in both methods the average ratio for some 90 predictions was close to unity, i.e. 1.05 and 0.92 respectively (Clayton *et al.*, 1988).

Rather than develop more complex settlement prediction methods, it is important to gather new case records of settlements. Typically, the settlement of a foundation on granular soil will be less than 25 mm. Taking the 90 settlement observations used by Clayton *et al.* (1988) and excluding narrow footings (where bearing capacity failure may have developed), the average settlement is only 13.9 mm. Only 10% of the settlement records had observed settlements greater than 25mm, and almost three quarters of these were for foundations wider than 10m. Of the remainder, 3 involved narrow foundations (1.2m breadth) loaded in excess of 750 kPa, and the remaining record had an *N*-value of 13 and a width of 6m.

It has been argued that, given the lack of accuracy of SPT-based methods of predicting the settlements of foundations in granular soil, it is important to identify, at an early stage, those situations where large settlements may occur, such as

- narrow, heavily stressed foundations
- very wide foundations
- foundations on metastable ('collapsing') soils, such as dune sands and lightly cemented sands (e.g. sabkha)
- sands containing small amounts or thin layers of organic or cohesive material.

For these situations, other test methods will be required, such as surface or borehole plate-loading tests, geophysical determination of shear-wave velocity, or self-boring pressuremeter tests.

9.2 DESIGN OF PILES IN SOILS, WEAK ROCKS AND CHALK

9.2.1 Suggested methods

(a) Shaft resistance

The estimation of available shaft resistance of piles is of great importance because in many soil conditions, at working loads, almost all of the load is transferred to the soil by this mechanism. Experience suggests that the ultimate shaft resistance of a pile will depend not only upon the ground conditions, but also upon its method of installation. Poulos (1989) has proposed the relationship

$$f_s = \alpha + \beta N$$

where f_s is the ultimate unit shaft resistance of the pile in kPa

N is SPT penetration resistance

α, β are constants depending upon soil and pile type.

Table 12 gives correlations between the shaft resistance of piles and the measured SPT N value of the soil, as proposed by Poulos, for preliminary pile design. The values given in the table should be used with the average penetration resistance over the relevant depth of the pile, in order to give the average ultimate shaft resistance per unit shaft area over that length.

The values of α and β depend to a considerable extent on pile type, because of pile installation effects. For example, for driven displacement piles, shaft resistance will depend upon the extent to which the pile displaces the soil, upon the rate at which excess pore pressures can dissipate during and after driving, and upon the amount of lateral movement allowed during driving. For cohesionless soils, Meyerhof (1956, 1965) suggests that the unit shaft resistance for large displacement piles will be approximately twice that for small-displacement (e.g. steel H) piles.

Table 12 gives and values for cohesive and cohesionless soils, and chalk. The values for cohesive soils based upon Japanese experience (Shioi and Fukui, 1982; Yamashita et al., 1987) seem relatively high by European experience and should be used with caution. Tomlinson (1965) suggests that, for driven piles in clay, unit shaft resistance is a function of soil conditions and pile length. Typically, the equivalent value would be 0, with β falling between 1.8 and 6.7. For bored cast-in-place piles in clay a value of between 2.7 and 3.3 would seem appropriate, based upon experience in the London Clay. In other, less well-known ground conditions an initial estimate (prior to pile load tests) may be based upon β = 2.0 to 2.3.

Table 12 Correlations between ultimate shaft resistance, fs, of piles and SPT N value (after Poulos, 1989)

Pile type	Soil type	α	β	Remarks	Reference
Driven displacement	Cohesionless	0	2.0	halve f_s, for small displacement	Meyerhof (1956) Shioi and Fukui (1982)
	Cohesionless and cohesive	10	3.3	pile type not specified $50 \geq N \geq 3$ $f_s \ngtr 170$ kPa	Decourt (1982)
	Cohesive	0	10*	–	Shioi and Fukui (1982)
Driven cast-in-place	Cohesionless	30	2.0	$f_s \ngtr 200$ kPa	Yamashita et al. (1987)
		0	5.0		Shioi and Fukui (1982)
	Cohesive	0	5.0*	$f_s \ngtr 150$ kPa	Yamashita et al. (1987)
		0	10.0*		Shioi and Fukui (1982)
Bored	Cohesionless	0	1.0		Findlay (1984) Shioi and Fukui (1982)
		0	3.3		Wright and Reese (1979)
	Cohesive	0	5.0*		Shioi and Fukui (1982)
		10	3.3	piles cast under bentonite $0 \geq N \geq$ $f_s \ngtr 170$ kPa	Decourt (1982)
	Chalk	–125	12.5	$30 \geq N \geq 15$ $f_s \ngtr 250$ kPa	Fletcher and Mizon (1984)

$f_s = \alpha + \beta.N$ (kPa)   *see note in text

Table 13 gives the method for assessing the ultimate shaft adhesion of driven piles in chalk proposed by Hobbs and Healy (1979). Once again, the method of pile installation has a significant effect on the available pile shaft adhesion, as expressed by Hobbs and Healy in terms of a varying compaction factor (C_f), and angle of friction between the pile wall and the chalk (δ'). The categories of piles used by Hobbs and Healy are:

1. Driven preformed piles.

2. Driven cast-in-place piles, where the *in-situ* concrete does not come directly in contact with the chalk.

3. Driven cast-in-place piles, where the fresh concrete comes directly in contact with the chalk.
4. Bored cast-in-place piles.

For bored cast-in-place piles, Hobbs and Healy suggest that the ultimate shaft load

$$Q_s = \pi.d \sum q_s . dz$$

where d is the shaft diameter and q_s is the ultimate shaft resistance.

The shaft adhesion or bond between the concrete and chalk can be estimated on the basis of back-analysed bored piles, and the results of special laboratory tests, correlated with penetration resistance. Hobbs and Healy suggest that the relationship between q_s and N can be approximated by

$q_s = 0.28\ N^2$ (kPa) where q_s should not exceed 250 kPa.

A factor of safety of 1.5 should be applied to calculated values of ultimate shaft resistance.

Table 13 *Estimation of ultimate shaft load for driven piles in chalk (from Hobbs and Healy, 1979)*

Pile material	C_f	$\tan\delta$
Categories 1 and 2	(max. value of $K = 2$)	
Precast concrete	1	0.45
Steel shell	1	0.35
Category 3(a) (zero slump concrete)	3	0.65
	(max. value of $C_f K = 4$)	
Category 3(b) (high slump concrete)		
GKN cast-in-place	1.25	0.65
Holmpress (single redrive)	2	0.65
	(max. value of $C_f K = 3$)	

$Q_s = \pi.d.\ \Sigma p'_z.C_f(0.06\ N_z) \tan\delta.\Delta z$ (kN)

(b) Base resistance

Values of base resistance suggested by Poulos (1989) based upon SPT penetration resistance are given in Table 14. The ultimate base resistance of a pile is calculated from

$q_b = K.N$ (kPa)

Once again, the values given for piles in clay seem high. A maximum of $K = 100$ is suggested.

Table 14 *Correlations between end-bearing resistance, fb, of piles and SPT N value (after Poulos, 1989)*

Pile type	Soil type	K	Remarks	Reference
Driven displacement	Sand	450		Martin et al. (1987)
	Sand	400		Decourt (1982)
	Silt, sandy silt	350		Martin *et al.* (1987)
	Glacial coarse to fine silt deposits	250		Thorburn & Mac Vicar (1971)
	Residual sandy silts	250		Decourt (1982)
	Residual clayey silts	200		Decourt (1982)
	Clay	200*		Martin *et al.* (1987)
	Clay	120*		Decourt (1982)
	All soils	300	For L/d ≥ 5	Shioi & Fukui (1982)
		(0.1 + 40 L/d)	If L/d < 5, closed-end piles	
		(60 L/d)	open-ended piles	
Driven cast-in-place	Cohesionless		q_b = 3000 kPa	Shioi & Fukui (1982)
		150	q_b ≯ 7500 kPa	Yamashita *et al.* (1987)
	Cohesive	-	q_b = 90 (1 + 160z) z = tip depth (m)	Yamashita *et al.* (1987)
Bored	Sand	100		Shioi & Fukui (1982)
	Clay	150*		Shioi & Fukui (1982)
	Chalk	250	N < 30	Hobbs (1977)
		200	N > 40	

$q_b = K.N.$ (kPa)   *see note in text

Hobbs and Healy (1979) similarly express base loads for piles in chalk, and their suggested values are given in Table 15.

In the UK much of the data used to obtain mass strengths in weak rocks and in chalk (see Sections 8.3.1 and 8.4.1) have been derived from the back-analysis of piles. The various relationships can then be interpreted directly in terms of ultimate bearing capacity, q_b, and penetration resistance, N.

For clays:

q_b = 45 to 75 N_{60} (kPa)   (based on Stroud, 1974)

For weak rocks:

q_b = 45 to 180 N_{60} (kPa) (based upon Leach and Thompson, 1979; Cole and Stroud, 1976; and Stroud, 1989).

For chalk:

q_b = 225 N_{60} (kPa) – bored piles
q_b = 250 N_{60} (kPa) – driven piles
q_b = 300 N_{60} (kPa) – driven precast concrete piles (based upon Lord, 1990).

Table 15 *Estimation of ultimate base load for piles in chalk (from Hobbs and Healy, 1970)*

Lower-bound N for 2 diameters below toe	K
<30	240
30–40	240–4(N–30)
>40	200

$Q_b = (\pi.d^2/4).K.N$ (kN)

9.2.2 Discussion

The SPT can provide valuable estimates of pile shaft resistance, particularly in cohesive soils and most weak rocks. In ground and drilling conditions unfavourable to the SPT (for example, light percussion drilling in granular soils), other design methods should also be used, especially if the results of the analysis are critical. In particular, the shaft resistance in chalk can be estimated by more rigorous methods (see, for example, Montague, 1990 and Lord 1990), which should be used for final designs.

The behaviour of piles in end-bearing is rarely observed, since shaft adhesion must be fully mobilised before end-bearing is brought into play, and large displacements and loads may well be necessary to achieve ultimate loads. Estimates in the literature are probably not, therefore, of great accuracy. The same observation may be made about N values in weak rocks and the harder chalks, as these normally require extrapolation.

9.3 LIQUEFACTION POTENTIAL IN GRANULAR SOILS

(a) Suggested method

The SPT is a dynamic test in which liquefaction of surrounding soil has been observed (Clayton and Dikran, 1982). Perhaps not surprisingly, therefore, it has been found to be a very useful indicator of liquefaction resistance, and indeed is considered by many engineers as not only very much simpler and cheaper, but also more reliable than laboratory test methods.

The maximum acceleration at the ground surface that can be sustained without liquefaction or cyclic mobility (Seed *et al.*, 1983; Seed *et al.*, 1985; Tokimatsu, 1988) is determined from Figures 48 to 51.

The maximum cyclic stress ratio is:

$$\frac{\tau_{av}}{\sigma'_o} = 0.65 \frac{a_{max}}{g} \cdot \frac{\sigma_o}{\sigma'_o} \cdot r_d$$

where

τ_{av}	=	verage applied shear stress
σ'_o	=	effective overburden stress at the depth under consideration
σ_o	=	total overburden stress at the depth under consideration
α_{max}	=	maximum acceleration at ground surface
g	=	gravitational constant
r_d	=	a stress reduction factor, which decreases from 1 at the ground surface to about 0.9 at a depth of 10m.

The limiting cyclic stress ratio, above which liquefaction may be expected to occur, is a function of:

- the SPT N value, corrected to 60% energy and to an overburden pressure of 100 kN/m², but not taking into account lateral stress increase due to over-consolidation
- the fines content (i.e. material finer than sand size) of the soil
- the magnitude of the earthquake.

Figure 48 compares the limiting curves from Seed *et al.* (1985) and Tokimatsu and Yoshimi (1983), for a magnitude 7.5 earthquake and a clean sand deposit (5% or less fines). Figure 49 shows how the effect of increasing fines content changes this limiting curve; the limiting penetration resistance is smaller, presumably because of increased pore pressure generation during driving, at any given cyclic stress ratio. Figure 50 shows how increasing earthquake magnitude decreases the permissible cyclic stress ratio, for a given penetration resistance, and Figure 51 shows the expected shear strain levels, γ_1, for a magnitude 7.5 event, once liquefaction occurs.

(b) Discussion

The curves given in Figures 48 to 51 are based upon field evidence of liquefaction, including sand boils, ground cracking, small lateral ground movements and, of course, major settlement, translation and bearing capacity failure of different types of structures. The standard penetration procedure to be adopted when using these figures is recommended by Seed *et al.* (1985) to be:

(a) 100-125mm dia. rotary borehole with bentonite drilling mud.

(b) Upward deflected drilling mud at bit (tricone or baffled drag bit).

(c) Standard split spoon (34.5mm internal diameter, 50.8mm outside diameter - i.d. to be constant along length of barrel).

(d) A or AW drill rods to 15m, N or NW rods at greater depth.

(e) 60% free-fall energy.

(f) 30-40 blows per minute.

(g) Penetration resistance measured over penetration range 152-457mm.

The techniques described above are suitable only in sands. Although the liquefaction of gravelly soils is well-documented (Harder and Seed, 1986) the SPT will provide erroneously high blow counts once the D_{50} particle size exceeds about 0.5mm.

Once D_{50} exceeds about 3mm it will be necessary to use larger penetrometers, which are less affected by particle size, such as the LPT (Kaito *et al.*, 1971; Yoshida *et al.*, 1988; Tokimatsu, 1988; Crova *et al.*, 1992) or the Becker Drill Hammer (Harder and Seed, 1986; Harder, 1988). Below this value the tentative correction factors for particle size suggested by Tokimatsu (1988) may be of use (Figure 30b).

9.4 ESTIMATION OF SHEET PILE DRIVABILITY IN GRANULAR SOILS

Suggested method

Although there are well-developed theories to permit the design of anchored and sheet-pile retaining walls (for example see Clayton *et al.*, 1993), the construction of these structures also requires that the steel section be strong enough to resist driving into the soil without buckling and de-clutching. As the effective angle of friction of the ground increases, the size of the section required to resist applied bending moments is reduced; but the size of sheet piling necessary to allow it to be driven into the ground must be increased.

Little is known about the necessary sections to allow sheet piles to be driven without damage. However, Williams and Waite (1993) suggest that in granular soils the minimum section can be judged on the basis of SPT N value, as given in Table 16.

Table 16 *Preliminary selection of sheet pile section to suit driving conditions in granular soils (Williams and Waite, 1993)*

Dominant SPT (N value)	Minimum wall modulus cm³/m of wall		Remarks
	BSEN 10 025 grade 430A, BS 4360 grade 43A	BSEN 10 025 grade 510A, BS 4360 grade 50A	
0 – 10	450		Grade FE510A for lengths greater than 10 m
11 – 20		450	
21 – 25	850		
26 – 30		850	Lengths greater than 15 m not advisable
31 – 35	1300		Penetration of such a stratum greater than 5 m not advisable*
36 – 40		1300	Penetration of such a stratum greater than 8 m not advisable*
41 – 45	2300		
46 – 50		2300	
51 – 60	3000		
61 – 70		3000	Some declutching may occur
71 – 80	4200		Some declutching may occur with pile lengths greater than 15 m
81 – 140		4200	Increased risk of declutching. Some piles may refuse

* If the stratum is of greater thickness use a larger section of pile

Note: This table is based on sheet pile sections of approximately 500 mm interlock centres, installed with panel driving techniques. Wider sections, and those installed by methods giving less control will require greater minimum moduli.

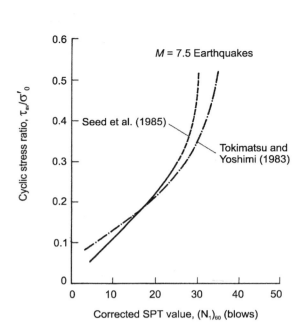

Figure 48 *Maximum cyclic stress ratio versus penetration resistance for a magnitude 7.5 earthquake in clean sand (after Tokimatsa, 1988)*

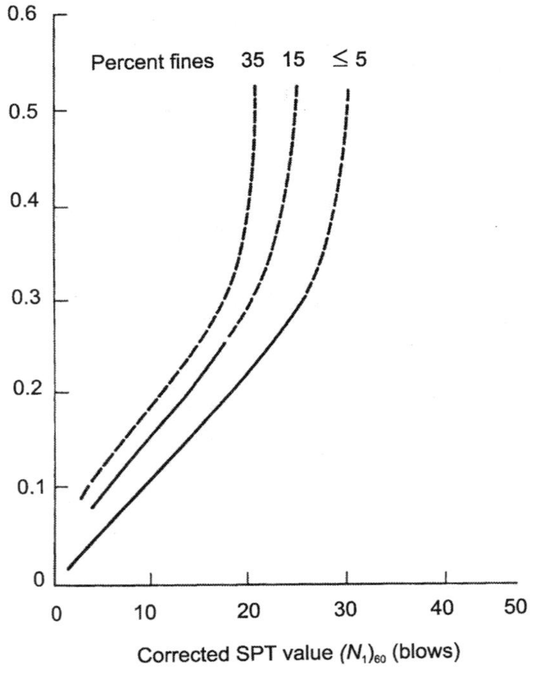

Figure 49 *Effect of fines content on limiting cyclic stress ratio for sand (after Seed et al., 1985)*

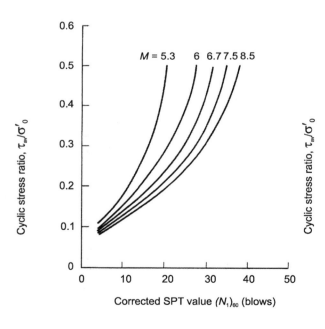

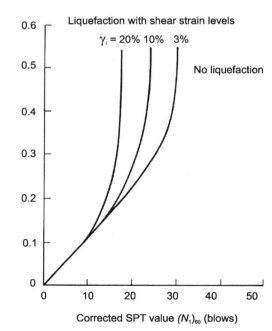

Figure 50 *Effect of earthquake magnitude, M, on limiting cyclic stress ratio (after Tokimatsu, 1988)*

Figure 51 *Expected shear strain levels following liquefaction of dense sand (after Seed et al., 1985)*

10 Recommendations for the use of the SPT

The SPT continues to be used in the UK and throughout the world, because it is valued as a simple indicator of the consistency of the ground, particularly in soils and weak rocks that are difficult to sample.

This widespread use, which at times can lead to complete reliance upon its results, is likely to continue for many years. It is important, therefore, that engineers specifying, supervising, reporting and using the results of the SPT should not only recognise its limitations, but should also try to achieve greater consistency in the test method and the use of its results. The recommendations given below relate not only to field procedures, but also to reporting and application of results, and the need for further studies.

10.1 SPT FIELD METHODS

Much of the required good practice for the SPT has been detailed in the current British Standard, and in the International Reference Test Procedure (IRTP) (Appendix 1). However, it has been noted that these documents remain, to some extent, incomplete. The following is intended to provide *recommendations* for the test in all ground conditions, and to give guidance on additional techniques, not detailed in the British Standard, which can be of help.

10.1.1 Boring and drilling techniques

- *In firm or stiff cohesive soils*, drilling disturbance is generally not great, whether using light percussion or rotary drilling techniques. For very soft cohesive soils, it is recommended that the hole should be maintained full of fluid (either water or mud) following Rowe (1972). In more competent cohesive soils, no special precautions are necessary. Neither drilling method, nor borehole diameter, appear critical.

- *In granular soils*, the boring or drilling method is known to have a significant influence on the ground disturbance, and hence on the SPT N value. The fundamental dilemma is that while it is difficult to bore or drill in soils containing coarse particles using small diameter holes, the use of large diameter holes renders the results of the SPT at best doubtful, and at worst seriously in error.

    BS 1377: 1990 suggests that N may be significantly affected when the borehole diameter used (in granular soils) is 150 mm or more, and that boring tools should have a diameter not greater than 90% of the inside diameter of the internal diameter of the casing. Boring tools have to be withdrawn slowly to minimise rates of upward groundwater flow into the base of the hole. Under normal conditions, the water or mud level in the borehole must, at all times, be maintained a sufficient distance above groundwater level to minimise disturbance at the base of the hole. When drilling in artesian groundwater conditions the casing must extend a sufficient height above ground level. The casing must not, at any time, be driven below the base of the borehole, into the test section.

It is now clear that *routine* light percussion techniques cannot be used in granular soils without causing serious loosening to at least 1½ and probably up to about 3 times the borehole or drillhole diameter. With light percussion drilling, therefore, the borehole diameter must not exceed 150 mm. If the SPT result is not to be affected by disturbance then the borehole diameter should not exceed 75 to 100mm.

If larger diameter boreholes are needed, perhaps because of the required depth of investigation or because of the presence of coarse (gravel or cobble) sized particles, then the method of drilling immediately above the SPT test section has to be changed. Casing should not be used within 1m of the base of the hole, in order that upflow of groundwater into the borehole is minimised. Therefore mud support will be required. Shells should not be used.

In finer soils a tricone or baffled drag bit, with drilling mud deflected upwards at the bit, will give the best quality results. In coarse granular soils an undersize short auger or barrel auger, lifted slowly from the hole, may be the only method of making progress without causing disturbance. Such a method will prove extremely costly in practice, and other *in situ* test methods should therefore be considered, particularly bearing in mind the fact that penetration resistance is significantly increased by the grading of such soils.

- In *weak rocks*, apart from chalk, no particular precautions appear to be necessary when drilling immediately above the test section. *In chalk*, however, it is clear that the detailed method of drilling has a significant effect on N. But no systematic study has yet been carried out in order to identify the mechanisms of disturbance. It is not possible, therefore, to make recommendations for the method of drilling to be used.

10.1.2 Test equipment

- The Standard Penetration Test must be carried out with an automatic trip hammer of the type similar to that sold in the UK by DANDO or EDECO. These hammers have a 62.5 kg mass which is tripped automatically 762 mm above an anvil with a mass of approximately 15-20 kg. The hammer must be maintained vertical during its use, in order to minimise frictional losses on the shaft.

- It is thought that a hammer of the type described above will consistently deliver to the rods between 70% and 75% of the free-fall energy of the falling weight. For most uses of the SPT data it will be sufficient to assume that this is the case, since other factors, such as drilling disturbance, are generally more critical. However, when precision is important, or when the recommended type of trip hammer is not available, energy measurements should be made (see Appendix 2).

- When working overseas it may not be possible to obtain automatic trip hammers of the varieties specified above. In this case the energy delivered by the local hammers and dropping procedures should be measured either in accordance with ASTM D4633-86 or by using the procedures described in Appendix 2. Energy measurements should aim not only to identify the differences in mean measured rod energy, attributable to differences in hammer/anvil design and method of tripping, but also the standard deviation for each hammer type and method of drop.

- The rods used to connect the trip hammer anvil to the split spoon must be AW (BS 4019) or stiffer. When testing at depths greater than 20m, BW rods, or stiffer, must be used. Rods heavier than 10 kg/m and bent rods, with a relative deflection of 1/1000 or greater, must not be used.

- The split spoon must comply with all the dimensions shown on Figure 4(b). (Note that the length of the shoe taper is incorrectly quoted in BS 1377: 1990, Part 9, and should be 19 mm, not 10 mm). A ball-check valve must be used, and the venting system maintained clean. Split-spoons with the provision for inclusion of liners are not recommended, nor is the use of catcher systems behind the shoe.

- The use of a 60° cone-end is permitted by BS 1377: 1990 in gravelly soils but is specifically excluded from the International Reference Test Procedure (IRTP). There is evidence, in sand, that the use of a solid cone in place of the normal open shoe can lead to considerably higher penetration resistances. Therefore it is recommended that the solid cone should only be used in granular soil if comparative tests in the particular ground conditions of the site have established that no increase in resistance is caused. The solid cone may be used in coarse gravelly soils, but its possible influence on N should be recognised. It should not be used in weak rocks.

10.1.3 Test frequency

In many types of ground, the SPT has poor repeatability. Typically, then, the mean value of the N v. depth profile is used in design. As with undrained triaxial tests, it is therefore necessary to make sufficient tests to have confidence in the mean. The SPT should be carried out at intervals of not more than 1.5m down the borehole. In near-surface deposits (depth less than 5m) a frequency of one test per metre is recommended.

10.1.4 Test procedure: normal circumstances

1. The depth to the bottom of the borehole should be measured immediately before the SPT tools are lowered into the hole, using a weighted tape. Once the borehole is found to be properly cleaned, the split spoon and rods can be lowered into the borehole, and the hammer attached. The depth of the bottom of the split spoon assembly must then be determined, from the lengths of the spoon and rods lowered into the hole and the measured stick-up, and compared with the depth to the base of the hole. If the bottom of the split spoon is found to be above the previously measured level of the bottom of the borehole, the assembly should be removed and the borehole re-cleaned. The penetration of the split-spoon under the weight of the rods and the trip hammer should be recorded, together with the depths of the borehole and casing.

2. Before the start of driving, the rods must be marked to show their level at the end of each of six drives of 75 mm. The seating drive can then commence through the first 150 mm of penetration. Within this length the blows required to drive the split-spoon through each increment of 75 mm must be recorded, separately.

3. The test drive is then started, which comprises a further 300 mm penetration. The blows required to drive the split-spoon through four increments of 75 mm must be recorded, separately.

4. At all times the blows should not be applied at a rate faster than 30 per minute.

5. Following withdrawal, the split spoon should be opened and the sample described and its length recorded. A representative portion of each soil type should be placed in a separate airtight container, and labelled with the site name, borehole number, and sample number.

10.1.5 Test procedure: hard ground

1. The seating drive may be terminated after 25 blows whether reached in either the first or second 75 mm of penetration and the total penetration recorded to the nearest 5 mm, in addition reporting the blow count for the first 75 mm of penetration, if this was achieved.

2. The test drive may be stopped at a limit of an additional 100 blows. The total penetration during the test drive should be reported to the nearest 5 mm. In addition, the blow count for each completed 75 mm increment must also be reported.

10.1.6 Suggested procedure for extending the drive

In granular soil, when the blow count for the test drive is low, erratic or less than expected, it has been found worthwhile to extend the test drive. This involves very little extra cost, and may assist in the identification of changes of strata and borehole disturbance or the presence of obstructions such as cobbles and flints. Suggested options are as follows:

(i) If the total blow count (seating drive + test drive) is less than 50, continue to a total penetration of 600 mm (i.e. 450mm test drive) or 80 blows, whichever occurs first.

(ii) If the total blow count (seating drive + test drive) is less than 20, continue to a total penetration of 750 mm (i.e. 600mm test drive) or 80 blows, whichever occurs first.

In both cases the blows for every 75 mm of penetration should be recorded.

10.1.7 Reporting

1. The record of the test must state the type of penetrometer used (i.e. SPT or SPT(C)), the date of the test, the borehole number and site name.

2. The basic steps taken in preparing for each test must be reported by the rig foreman on the daily record sheet, as:

   – method of boring/drilling

   – drilling tool (and its diameter) used to prepare the section of hole above the test section (this may be stated in the report when normal methods are in use)

   – borehole diameter and depth of borehole at start of test

   – casing diameter, and depth at time of test

   – drilling fluid type, and level at time of test

   – depth to bottom of split-spoon after trip hammer connection

   – blows per 75 mm, and total penetration, during seating drive

   – blows per 75 mm, and total penetration, during test drive.

3. The recorded details are to be given in a form similar to that shown in Table 17.

Table 17 *Examples of SPT records on rig foreman's daily reports*

Date	Measured borehole depth (m)	Casing depth/dia. (m)	Depth to water (m)	Test type	Test depth (m)	Result
26.6.91	10.58	3.05/0.20	dry	SPT	10.65-11.10	6/8/12 13/12/10
12.3.89	6.25	5.70/0.15	1.00	SPT	6.73-7.18	2/5/7 9/6/8
15.5.90	8.90	8.84/0.20	4.50	SPT	9.00-9.75	1/1/0/1/2 3/2/3/2/4
28.8.91	15.62	14.84/0.20	8.30	SPT(C)	15.60-16.05	3/5/7 8/7/9
12.1.90	23.50	7.00/0.20	20.50	SPT*	23.50-23.78	15/10-125 33/47/20-155

Location: Site A Borehole: 23A
Type of rig: Pilcon Wayfarer 1500
Method of drilling: light percussion

10.2 INTERPRETATION

1. Under normal conditions, where the total SPT penetration consists of the penetration of the split spoon under its self weight and the weight of the rods and hammer assembly, plus six 75 mm-long drives, the N value shall be the sum of the last four 75 mm-long drives.

2. The SPT resistance, N, shall be reported on the engineer's borehole records, together with the blow count for each increment of both the seating drive and the test drive. Where a part of the drive is short, its length shall also be reported. The true depth of the bottom of the split spoon at the start of the test (i.e. the depth of the borehole plus the measured penetration under the dead weight of the test apparatus) shall be reported, together with the depth at the end of the drive. Examples are given in Table 18.

3. N values given on borehole records should not include corrections for overburden pressure (e.g. to give N_1) or energy (e.g. to give N_{60}). Density descriptors given in the engineering sample description of granular soils should reflect the expected state of density of the soil, and should therefore be based upon calculated $(N_1)_{60}$ values, using the classification system given in Table 8. Corrections for borehole diameter and short rod lengths are not recommended.

4. When high driving resistance is encountered, for example in rock, a normal penetration resistance cannot be achieved. The extrapolated N value, N^*, shall be calculated from

$$N^* = \frac{\text{blows during test drive (normally 100)}}{\text{penetration during test drive (mm)}} \times 300$$

5. When very low driving resistance is encountered, and the SPT assembly sinks by more than 450 mm under its self weight, the British Standard suggests that an N value of 0 should be quoted. However, the results of good quality SPT testing suggest that such an occurrence will be because of poor drilling techniques. Therefore, in contrast, it is here recommended that the N value should continue to be the sum of the blows for the penetration from 150 to 450 mm *beyond* the initial self-weight penetration.

Table 18 *Examples of SPT records on engineering report of borehole record*

Date	Test type	Test depth (m)	Result	N	Comments
26.6.91	SPT	10.65–11.10	6/8/12 13/12/10	47	Normal SPT, in clay
12.3.89	SPT	6.73–7.18	2/5/7 9/6/8	30	SPT with high initial penetration under self-weight
15.5.90	SPT	9.00–9.75	1/1/0/1/2 3/2/3/2/4	6	SPT with extended drive – suspected borehole disturbance of granular soil
28.8.91	SPT(C)	15.60–16.05	3/5/7 8/7/9	31	SPT with solid cone, in gravel
12.1.90	SPT*	23.50–23.78	15/10-125 33/47/20-105	286	SPT in weak rock, with short drive

Location: Site A Borehole: 23A
Type of rig: Pilcon Wayfarer 1500
Method of drilling: light percussion

6. If the suggested procedure for extending the drive in loose granular soils is adopted, the reported N value should still be as defined above, i.e. the blow count between 150 and 450 mm of penetration beyond the seating drive.

7. Where there are doubts about the validity of the penetration resistance values, as calculated in the routine way, the engineer responsible for the geotechnical design of the project should consider reinterpreting these values. Any reinterpreted values should be reported *in addition* to the routine values, and the reasons for the perceived need to reinterpret the data should be clearly stated in the engineering report. Reinterpreted values should *not* be given on the engineering borehole records, or in factual ground investigation reports.

8. Under normal circumstances, the blow counts for each 75 mm will show a regular rate of increase with penetration. When a distinct change in the rate of increase in blows/ increment occurs it may indicate:

    – depth of disturbance greater than 150 mm

    – the presence of an obstruction, such as a cobble or a flint

    – a change of strata

    – anomalous ground, such as a thin clay layer in granular soil, or a weak or fractured zone in marl or shale.

    When such conditions are suspected, and the SPT records are variable, reinterpretation of individual N values using engineering judgement may help to reduce the scatter of penetration resistance values on a depth profile. In addition, such results may indicate the necessity for additional investigation of the site using other methods, for example the cone penetration test (CPT).

10.3 SUPERVISION AND TRAINING

The N value obtained from the SPT is a function, in many soil types, of the way in which the borehole is made, and the care with which data are recorded. Thus supervision and training are most important, not only to ensure that the work meets the requirements of the job and complies with current good practice, but also to help the driller appreciate how much the quality of the ground investigation depends upon his care.

Under ideal circumstances, drilling should be supervised at all times, preferably by a specialist drilling technician who could control not only the testing and sampling procedures, but also carry out logging and sample description. While such a high level of supervision is used by some geotechnical companies in the USA, it is rare in the UK. If non-specialist supervisors (for example, engineering graduates) are to be used on site then it is recommended that they receive training in ground investigation techniques beforehand. On critical jobs, supervising engineers should be persons listed in the British Geotechnical Society's Directory of Geotechnical Specialists.

With regard to driller training, the situation in the United Kingdom has recently improved, with the implementation of the British Drilling Association (BDA) Driller Accreditation Scheme, which is designed to improve the standards of ground investigation fieldwork. Nevertheless, there is still a need for specific training courses to instruct accredited drillers in detail on all aspects of the SPT, including:

– drilling techniques

– testing and reporting

– uses of the SPT in engineering design.

10.4 APPLICATION OF SPT TEST RESULTS

The SPT is a common test. Its great strengths are its simplicity, its low cost, and its wide applicability. However, its weaknesses include lack of standardisation, and the large influence that drilling techniques may have, albeit in a limited range of ground conditions.

Sections 7, 8 and 9 detail the uses of the SPT. Broadly, they can be classified as profiling, classification, parameter determination, and direct design. In all of these applications the particular problems related to the specific ground conditions in which the test is carried out should be borne in mind:

1. In granular soils and chalk, the test result may be significantly influenced by disturbance caused by the method of forming the hole. Engineering design may be of limited accuracy.

2. Correlations between penetration resistance and the compressibility of granular soil will be of limited accuracy, because of the differences in the factors influencing them. As the D_{50} size of the soil exceeds 0.3mm, correlations developed for sands can no longer be relied upon.

3. When quoting SPT results in publications, for example in conference papers and engineering journals, all the elements of the IRTP should be clearly identified and, indeed, should have been followed during field testing.

4. In cohesive soils, provided that they are not sensitive, SPT results should be relatively reliable. In normally and lightly overconsolidated soils at shallow depth the test will be too insensitive to be of practical use. In over-consolidated fissured cohesive soils the test can provide a good guide to (100mm triaxial) undrained strength. But it is important to remember that undrained shear strength is a function of both test size and test method. The SPT sample should not be used for laboratory undisturbed strength tests as, at best, it is only Class 3 quality (BS 5930: 1981).

5. In weak rocks the SPT N^* value is useful, because there are no cheap alternatives. However, small penetrations often into variable lithologies mean that extrapolated values have little accuracy.

In general, the SPT may be used to obtain preliminary, conservative, estimates of engineering behaviour: in this, it is very effective, but it is good practice to check them against other available evidence. Where more precision is required, other techniques, either in situ or in the laboratory, should be used.

10.5 FURTHER WORK

As the SPT is still in such widespread use, it is vital that further work be carried out to identify and quantify both the strengths and limitations of the technique, and its uses and interpretation. Despite about 40 years of continuous SPT use in the UK, little is known about many aspects of its behaviour. In particular, further research is required on:

- the influence of particle size on penetration resistance
- the mechanisms of drilling disturbance on the chalk
- the energy delivered by British automatic trip hammers
- the effect of the solid cone on penetration resistance
- the influence of borehole diameter
- the extent of borehole disturbance in granular soils, and methods of reducing it
- the sensitivity of the extrapolated N value, obtained in weak rocks, to the method of test.

References

AMERICAN SOCIETY FOR TESTING AND MATERIALS (ASTM)
ASTM. D1586-58 (1958) *Standard method for Penetration Test and Split Barrel Sampling of Soils*
American Society for Testing and Materials, Philadelphia, USA

ASTM. D1586-67 (re-approved 1974)
Standard method for penetration test and split-barrel sampling of soils
American Society for Testing and Materials, Philadelphia, USA

ASTM. D1586-84 (re-approved 1992)
Standard test method for penetration test and split-barrel sampling of soils
American Society for Testing and Materials, Philadelphia, USA

ASTM. D4633-86 (1986) *Standard Test Method for Stress Wave Energy Measurement for Dynamic Penetrometer Testing Systems.*
American Society for Testing and Materials, Philadelphia, USA

BARTON, M.E., COOPER, M.R. and PALMER, S.N. (1989)
Diagenetic alteration and micro-structural characteristics of sands: neglected factors in the interpretation of penetration tests
Proc. ICE Conf. on Penetration Testing in the UK, Birmingham, Thomas Telford, London, 7-10

BAZARAA, A.R.S.S. (1967)
Use of the standard penetration test for estimating settlement of shallow foundations on sand
Unpublished PhD Thesis, University. of Illinois, USA

BEGEMANN, H.K.S. Ph. and DE LEEUW, E.H. (1979)
Current Practice of the Sampling of Sandy Soils in the Netherlands. State-of-the-Art Report,
Proc. Int. Symp. of Soil Sampling, Singapore, Ed. ISSMFE Sub-committee on Soil Sampling. Jap. Soc. SMFE., pp **55-6**

BICA, A.V.D. and CLAYTON, C.R.I. (1989)
Design methods for free embedded cantilever walls
Proc. ICE, **86**, Oct., 879-98

BOLTON, M.D. (1986)
The strength and dilatancy of sands
Geotechnique, **36**, 1, 65-78

BRATCHELL, G.E., LEGGATT, A.J. and SIMONS, N.E. (1975)
The performance of two large oil tanks founded on compacted gravel at Fawley, Southampton, Hampshire
Proc. BGS Conf. on Settlement of Structures, Cambridge, 3-9. Pentech Press, London

BRITISH STANDARDS INSTITUTION
BS 1377:1967, *Methods of testing soils for civil engineering purposes*
British Standards Institution, London

BS 1377: 1975, *Methods of test for soils for civil engineering purposes*
British Standards Institution, London

BS 1377: Part 9:1990, British Standard *Methods of test for soils for civil engineering purposes. Part 9. In situ tests*
British Standards Institution, London

BS 4019: Part 1:1974, *Specification for rotary core drilling equipment. Part 1. Basic equipment*
British Standards Institution, London

BROWN, R.E. (1977)
Drill rod influence on Standard Penetration Test
Proc. ASCE, Jn. Geot. Engg Div., GT11, 103, 1322-36

BURBIDGE, M.C. (1982)
A case study review of settlements on granular soil
Unpublished MSc Dissertation, Imperial College, London University

BURLAND, J.B., BROMS, B.B. and DE MELLO, V.F.B. (1977)
Behaviour of foundations and structures, State-of-the-Art review
Proc. 9th Int. Conf. Soil Mech. Found. Engg., Tokyo, 3, 395-546

BURLAND, J.B. and BURBIDGE, M.C. (1985)
Settlement of foundations on sand and gravel
Proc. ICE, Part 1, **78**, 1325-71

BURLAND, J.B. and LORD, J.A. (1970)
The load-deformation behaviour of Middle Chalk at Mundford, Norfolk; a comparison between full-scale and laboratory measurements
Proc. Conf. on In Situ Investigations in Soil and Rock, BGS, London

BUTLER, F.G. (1975)
Heavily overconsolidated clays
General Report and State-of-the-Art review for Session 3.
Proc. Conf. on Settlement of Structures, Cambridge. Pentech Press, London

CANADIAN STANDARDS ASSOCIATION
CSA A119.1 (1966) *Code for Split-barrel Sampling of Soils*
Canadian Standards Association, Ottawa

CHRISTOULAS, S. and PACHAKIS, M. (1987)
Pile settlement based on SPT results
Bull. Res. Centre of Public Works Dept. of Greece, **3**, July-Sept., 221-6

CLAYTON, C.R.I. (1978)
A note on the effects of density on the results of Standard Penetration Tests in Chalk
Geotechnique, **28**, 1, 119-22

CLAYTON, C.R.I. (1986)
Discussion on: The settlement of foundations on granular soils by Burland and Burbidge (1986)
Proc. ICE, **80**, 1, 1630-33

CLAYTON, C.R.I. (1990) (a)
The mechanical properties of the Chalk
Keynote Address. *Proc. Int. Chalk Symp.*, Brighton, 213-32. Telford, London

CLAYTON, C.R.I. (1990) (b)
SPT energy transmission: theory measurement and significance
Ground Engg, **23**, 10, 33-42

CLAYTON, C.R.I. and DIKRAN, S.S. (1982)
Porewater pressures generated during dynamic penetration testing
Proc. 2nd Eur. Symp. on Penetration Testing (ESOPT-II), Amsterdam, Vol. 2, pp 245-50

CLAYTON, C.R.I., HABABA, M.B. and SIMONS, N.E. (1985)
Dynamic penetration resistance and the prediction of the compressibility of a fine sand - a laboratory study. *Geotechnique*, **35**, 1, 19-31

CLAYTON, C.R.I., MILITITSKY, J.M. and WOODS, R.I. (1993)
Earth Pressure and Earth-retaining Structures
Surrey University Press, Glasgow

CLAYTON, C.R.I., SIMONS, N.E. and INSTONE, S.J. (1988)
Research on dynamic penetration testing of sands
Proc. ISOPT - 1, Vol. 1, pp 415-22

CLAYTON, C.R.I., SIMONS, N.E. and MATTHEWS, M.C. (1983)
Site Investigation
Granada, London. 424 pp

COLE, K.W. and STROUD, M.A. (1976)
Rock socket piles at Coventry Point, Market Way, Coventry
Symp. on Piles in Weak Rock. *Geotechnique*, **26**, 1, 47-62

CONNOR, I.G. (1980)
A Study of Soil-structure Interaction with Particular Reference to Sugar Silos
Unpublished MSc Dissertation, University of Surrey

CORNFORTH, D.H. (1973)
Prediction of drained strength of sands from relative density measurements
Proc. Symp. Evaluation of Relative Density and its Role in Geotechnical Projects Involving Cohesionless Soil. ASTM STP 523, 281-303

CRESPELLANI, T. and VANNUCCHI, G. (1991)
Dynamic properties of soils
In: *Seismic hazard and site effects in the Florence area* (ed. G. Vannucchi), 71-80.
Assoc. Geot. Italiana, Rome

CROVA, R., JAMIOLKOWSKI, M.B., LANCELLOTTA, R., and LO PRESTI, D.C.F. (1992)
Geotechnical characterization of gravelly soils at Messina Site - selected topics
Proc. Wroth Memorial Symposium on Predictive Soil Mechanics, Oxford, 27-29th July, 1992

CSN 731 821 (1961)
Stanoveni ulehlasti pisku dynamickou penetracni zkouskou
(Determination of sand density by the dynamic penetration test)
Czechoslovakian Standard CSN 73, 19 pp. Vydavatelstvi UN, Praha
(Publishing house of Office of Standardisation, Prague)

D'APPOLONIA, D.J. and D'APPOLONIA, E. (1970)
Use of the SPT to estimate settlement of footings on sand
Proc. Symp. on Foundations in Interbedded Sands, Perth, 16

D'APPOLONIA, D.J., D'APPOLONIA, E., and BRISETTE, R.F. (1970)
Closure to discussion: Settlement of spread footings on sand
Proc. ASCE, Jn. Soil Mech. Found. Div., **96**, SM2, 754-61

DAUNCEY, P.C. and WOODLAND, A.R. (1988)
Bored cast in situ piled foundations in Keuper Marl for the Birmingham International Arena
Proc. Conf. on Piling and Ground Treatment, Thomas Telford, London

DECOURT, L. (1982)
Prediction of the bearing capacity of piles based exclusively on N values of the SPT
Proc. 2nd Eur. Symp. on Penetration Testing (ESOPT II), Amsterdam, 1, 29-34

DECOURT, L. (1990)
The Standard Penetration Test, State-of-the-Art Report
Proc. 12th Int. Conf. on Soil Mech. and Found. Engg., Rio de Janerio

DECOURT, L. (1991)
Special problems on foundations
General Report, *Proc. IX Panamerican Conf. on Soil Mechanics and Foundation Engineering*, Vina del Mare, Chile

DENNEHY, J.P. (1975)
Correlating the SPT N value with chalk grade for some zones of the Upper Chalk
Geotechnique, **25**, 3, 610-14

DENVER, H. (1982)
Modulus of elasticity determined by SPT and CPT
Proc. 2nd Eur. Symp. on Penetration Testing (ESOPT II), Amsterdam, 1, 35-40

DE MELLO, V.F.B. (1971)
The Standard Penetration Test
State-of-the-Art Report. *4th Pan-American Conf. on Soil Mechanics Foundation Engineering*, Puerto Rico, 1, 1-86

DIKRAN, S.S. (1983)
Some factors affecting the dynamic penetration resistance of a saturated fine sand
Unpublished PhD Thesis, University of Surrey

DIN 4094, Part 2 (1980)
Dynamic and Static Penetrometers. Application and Evaluation
Deutsches Institut for Normung, Berlin

FINDLAY, J.D. (1984)
Discussion.
Proc. ICE Conf. on Piling and Ground Treatment, 189-90. Thomas Telford, London

FLETCHER, G.F.A. (1965)
Standard Penetration Test: its uses and abuses
Proc. ASCE, Jn. Soil Mech. Found. Div., **91**, SM4, 67-75

FLETCHER, M.S. and MIZON, D.H. (1984)
Piles in Chalk for Orwell Bridge
Proc. ICE Conf. on Piling and Ground Treatment, 203-9. Thomas Telford, London

GEISSER, R.F. (1966)
Discussion on Fletcher (1965)
Proc. ASCE, Jn. Soil Mech. Found. Div., **92**, SM2, 184-5

GIBBS, H.J. and HOLTZ, W.G. (1957)
Research on determining the density of sands by spoon penetration testing
Proc. 4th Int. Conf. Soil Mech. Found. Engg., London, 1, 35-9

GREENWOOD, D.A. (1986)
Private Communication about SPTs in relation to borehole size at Cairo Waste Water Project

HARDER, L.F. (1988)
Use of penetration tests to determine the cyclic loading resistance of gravelly soils during earthquake shaking
Unpublished PhD Thesis, University of California, Berkeley

HARDER, L.F. and SEED, H.B. (1986)
Determination of penetration resistance for coarse-grained soils using the Becker hammer drill
Report No. EERC - 86/06. Earthquake Engineering Research Centre

HARDING, H.J.B. (1949)
Site investigations including boring and other methods of sub-surface explorations
J. Instn. Civ. Eng., **32**, 111-37

HIEDRA-COBO, J.C. (1987)
Personal communication about Venezuelan practice

HOBBS, N.B. (1977)
Behaviour and design of piles in chalk – an introduction to the discussion of the papers on chalk
Piles in Weak Rock, 149-175. ICE, London

HOBBS, N.B. and HEALY, P.R. (1979)
Piling in Chalk
DOE/CIRIA Report PG6. CIRIA, London

HOLUBEC, I. and D'APPOLONIA, E. (1973)
Effect of particle shape on the engineering properties of granular soils
In: *ASTM STP 523*, 304-18

HVORSLEV, M.J. (1949)
Subsurface exploration and sampling of soils for civil engineering purposes
Waterways Experimental Station, Vicksburg, USA. 521 pp

IMAI, T. and TANOUCHI, K. (1982)
Correlation of N value with S-wave velocity and shear modulus
Proc. 2nd Eur. Symp. on Penetration Testing (ESOPT II), Amsterdam, Vol. 1

IMAI, T. and YOKOTA, K. (1982)
Relationships between N value and dynamic soil properties
Proc. 2nd Eur. Symp. on Penetration Testing (ESOPT II), Amsterdam, Vol. 1, pp 73-8

INDIAN STANDARDS INSTITUTION IS: 2131 (1981)
Method for Standard Penetration Test for Soils
Indian Standards Institution, New Delhi

INTERNATIONAL SOCIETY FOR SOIL MECHANICS AND FOUNDATION ENGINEERING (ISSMFE) (1977)
Report of the sub-committee on the penetration test for use in Europe
International Society for Soil Mechanics and Foundation Engineering, London

ISSMFE (1988)
Standard Penetration Test (SPT): International Reference Test Procedure
Proc. ISOPT-1, 1, 3-26

IRELAND, H.O. (1966)
Discussion on Fletcher (1965)
Proc. ASCE, Jn. Soil Mech. Found. Div., **92**, SM2, 189-190

IRELAND, H.O., MORETTO, O. and VARGAS, M. (1970)
The dynamic penetration test: a standard that is not standardized
Geotechnique, **20**, 2, 185-192

JAMIOLKOWSKI, M., BALDI, G., BELLOTTI, R., GHIONNA, V. and PASQUALINI, E. (1985)
Penetration resistance and liquefaction of sands
Proc. 11th Int. Conf. Soil Mech. Found. Engg., San Francisco, 4, 1891-6

JAPANESE INDUSTRY STANDARD (JIS) A-1219 (1976)
Method of penetration test for soils

JAPANESE SOCIETY OF SOIL MECHANICS (1983)
Standard Penetration Testing in Japan
Report of the Research Sub-committee on Site Investigation, Jap. Soc. Soil Mech. Found. Engng, to the Sub-committee on Penetration Testing, ISSMFE, at Helsinki.

JARDINE, F.M. (1986)
Discussion on Field Testing: The Standard Penetration Test, In: *Site Investigation Practice: Assessing BS 5930*, Geological Society, London (ed. Hawkins, A.B.), 27-29

JARDINE, F.M. (1989)
Standard penetration testing in the UK
Proc. Conf. on Penetration Testing in the UK, Birmingham. Thomas Telford, London

JAYAPALAN, J.K. and BOEHM, R. (1986)
Procedures for predicting settlements in sands
Proc. ASCE Conf., Seattle. Geotech. Spec. Pub. 5, 1-23

KAITO, T., SAKAGUCHI, S., NISHIGAKI, Y., MIKI, K. and YUKAMI, H. (1971)
Large penetration test
Tsuchi-to-Kiso, **629**, 15-21 (Japan)

KILBOURN, N.S., TREHARNE, G. and ZARIFIAN, V. (1989)
The use of the Standard Penetration Test for the design of bored piles in the Keuper Marl of Cardiff
Proc. Conf. on Penetration Testing in the UK, Birmingham. Thomas Telford, London

KOMORNIK, A. (1974)
Penetration testing in Israel
Proc. 1st Eur. Symp. on Penetration Testing (ESOPT I), Vol. 1, 185-92

KOREEDA, T., YOSHIHASHI, T. and MUROMACHI, T. (1980)
Comparing SPT N-value with JIS standard rod and another rod
Proc. Sounding Symp., Jap. Soc. Soil Mech. Found. Engg., 53-8 (in Japanese)

KOVACS, W.D., EVANS, J.C. and GRIFFITH, A.H. (1977)
Towards a more standardised SPT
Proc. 9th Int. Conf. Soil Mechanics and Foundation Engineering, Tokyo, **2**, 269-76

KOVACS, W.D. and SALOMONE, L.A. (1982)
SPT Hammer Energy Measurement
Jn. Geot. Engg. Div., Proc. ASCE, GT7, **108**, pp (No. 974)

KOVACS, W.D. and SALOMONE, L.A. (1984)
Field evaluation of SPT energy equipment, and methods in Japan compared with the SPT in the United States
US Dept. of Commerce, National Bureau of Standards, National Engineering Technology Center for Building Technology, Gaithersburg, USA,
Report NBSIR 84-2910

KOVACS, W.D., SALAMONE, L.A. and YOKEL, F.Y. (1982)
Energy measurement in the Standard Penetration Test
NBS Building Science Series 135, National Bureau of Standards, Washington, DC, Aug.

LACROIX, Y. and HORN, H.M. (1973)
Direct determination and indirect evaluation of relative density and its use on earthworks construction projects, In: ASTM STP 523, 251-80

LAKE, L.M. (1975)
Discussion on Session I: Granular Materials
Proc. BGS Conf. on Settlement of Structures, 663 Pentech Press, London

LEACH, B.A. and THOMPSON, R.P. (1979)
The design and performance of large diameter bored piles in weak mudstone rocks
Proc. 7th Eur. Conf. Soil Mechanics and Foundation Engineering., Brighton 1979, British Geotechnical Society, Vol. 3, 101-8

LE GRAND, SUTCLIFFE and GELL (1934)
Tool for producing true cores of clay
Engineering, (London), **138**, 23

LIAO, S. and WHITMAN, R.V. (1985)
Overburden correction factors for SPT in sand
Proc. ASCE, Jn. Geot. Engg., Div., 112, 3, 373-7

LONGWORTH, T.I. (1978)
Laboratory investigation of the local deformation properties of Middle Chalk
BRE Note N 74/78. Dept. of the Environment

LORD, J.A. (1990)
Foundations in Chalk
Keynote Address.
Proc. Int. Chalk Symp., Brighton, 301-26.
Telford, London

MCLEAN, F.G., FRANKLIN, A.G. and DAHLSTRAND, A.M. (1975)
Influence of mechanical variables on the SPT.
Proc. Conf. on In Situ Measurement of Soil Properties, Raleigh, N. Carolina, Vol. 1, pp 287-318

MALLARD, D.J. (1977)
Discussion: Session 1 - Chalk.
Proc. ICE Conf. on Piles in Weak Rock, pp 177-80

MALLARD, D.J. (1983)
Testing for Liquefaction Potential
Proc. NATO Workshop on Seismicity and Seismic Risk in the Offshore North Sea Area, Utrecht, 289-302, Reidel, Dordrecht

MARCUSON, W.F. and BIEGANOUSKY, W.A. (1977)
Laboratory Standard Penetration Tests on Fine Sands.
Jn. Geot. Engg, Div., Vol. 103, GT6, 565-80

MARSLAND, A. (1972)
Clays subjected to in situ plate tests
Ground Engg, **5**, 6, 24-31

MARTIN, R.E., SELI, J.J., POWELL, G.W. and BERTOULIN, M. (1987)
Concrete pile design in Tidewater, Virginia
Proc. ASCE, Jn. Geot. Engg. Div., **113**, 6, 568-85

MATSUMOTO, K. and MATSUBARA, M. (1982)
Effects of rod diameter in the Standard Penetration Test
Proc. 2nd. Eur. Symp. on Penetration Testing (ESOPT II), Amsterdam, 1, 107-12

MATTHEWS, M.C.G., CLAYTON, C.R.I. and RUSSELL, C.S. (1990)
Assessing the mass compressibility of Chalk from visual description
Proc. Conf. on the Engineering Geology of Weak Rock, Leeds

MEIGH, A.C. (1980)
Discussion
Proc. 7th Eur. Conf. Soil Mech. Found. Engg, Brighton, 1979, Vol. 4, pp 67-72

MEIGH, A.C. and NIXON, I.K. (1961)
Comparison of in situ tests for granular soils
Proc. 5th Int. Conf. Soil Mech. Found. Engg, Paris, Vol. 1, pp 499

MEYERHOF, G.G. (1956)
Penetration tests and bearing capacity of cohesionless soils.
J. Soil Mech. and Found. Div., Proc. ASCE, **82**, Jan., paper 866

MEYERHOF, G.G. (1957)
Discussion on Research on determining the density of sands by spoon penetration testing
Proc. 4th Int. Conf. Soil Mechanics and Foundation Engineering, London, Vol. 3, p110

MEYERHOF, G.G. (1965)
Shallow foundations
Proc. ASCE, Jn. Soil Mech. Found. Div., **91**, SM2, 21-31

MILITITSKY, J., CLAYTON, C.R.I., TALBOT, J.C.S. and DIKRAN, S. (1982)
Previsao de recalques em solas granulares ulitizando resultados de SPT: revisao critica
Proc. 7th Brazilian Conf. Soil Mechanics and Foundation Engineering, pp 133-50

MICHELL, J.K., GUZIKOWSKI, F. and VILLET, W.C.B. (1978)
The measurement of soil properties in situ, present methods – their applicability and potential
US Dept. of Energy Report, Dept. of Civil Engineering, University. of California, Berkeley

MOHR, H.A. (1937)
Exploration of Soil Conditions and Sampling Operations
Harvard Soil Mechanics Series No. 21

MOHR, H.A. (1966)
Discussion on Fletcher (1965)
Proc. ASCE, Jn. Soil Mech. Found. Div., **92**, SM1, 196-9

MONTAGUE, K.N. (1990)
SPT and pile performance in Upper Chalk
Proc. Int. Chalk Symp., Brighton Telford, London, 269-76.

MUROMACHI, T., OGURO, I. and MIYASHITA, T. (1974)
Penetration testing in Japan
Proc. 1st Eur. Symp. on Penetration Testing (ESOPT 1), Vol. 1, pp 193-200

NAISMITH, H.W. (1986)
Suit is a four letter word: a geotechnical engineer's introduction to professional liability
BiTech Publishers Ltd., Vancouver

NIXON, I.K. (1954)
Some investigations on granular soils with particular reference to the compressed-air sand sampler
Geotechnique, **4**, 16-31

NORMA BRASILEIRA REGISTRADA
NBR 6484-1980
Execuçao de Sondagens de simples reconhecimento dos solos
Norma Brasileira Registrada. Associaçao Brasileira de Normas Tecnicas

OHTA, Y. and GOTO, N. (1978)
Empirical shear wave velocity equations in terms of characteristic soil indexes
Earth Engineering and Structural Dynamics, Vol. 6, pp 61-73

OHYA, S., IMAI, T. and MATSUBARA, M. (1982)
Relationship between N – value by SPT and LLT measurement results
Proc. 2nd Eur. Symp. on Penetration Testing (ESOPT II), Amsterdam, 125-30

PALMER, D.J. and STUART, J.G. (1957)
Some observations on the Standard penetration test and a correlation of the test with a new penetrometer
Proc. 4th Int. Conf. Soil Mechanics and Foundation Engineering, London, 1, 231-6

PECK, R.B., HANSON, W.F. and THORNBURN, T.H. (1953)
Foundation Engineering
Wiley, New York (1st edn)

PECK, R.B., HANSON, W.E. and THORNBURN, T.H. (1974)
Foundation Engineering
Wiley, New York (2nd Edn)

POULOS, H.G. (1989)
Pile behaviour – theory and application
29th Rankine Lecture
Geotechnique, **39**, 3, 363-416

RANZINI, S. (1988)
SPTF
Technical Note, *Solos e Rochas* (Brazil), Vol. 11, pp 29-30 (in Portuguese)

REIDEL, W. (1929)
Das Aufquellen geologisher Schmelzmassen als plastischer Formanderungsvorgang
Neues Jahrbuck der Mineralogie, Beilagebande B, 62, 151-70 (see Hvorslev, M.J., 1949)

RIGGS, C.O. (1986)
North American Standard Penetration Test Practice - an essay
Proc. ASCE Conf. on use of in situ Tests in Geotechnical Engineering, Blacksburg, VA, 949-65

RIGGS, C.O., SCHMIDT, N.O. and RASSIEUR, C.L. (1983)
Reproducible SPT hammer impact force with an automatic free fall SPT hammer system
Geotechnical Testing J., **6**, 3, 201-9

RODIN, S., CORBETT, B.O., SHERWOOD, D.E. and THORBURN, S. (1974)
Penetration testing in United Kingdom,
Proc. 1st Eur. Symp. on Penetration Testing (ESOPT 1), Vol. 1, 139-46

ROWE, P.W. (1972)
The relevance of soil fabric to site investigation practice 12th Rankine Lecture
Geotechnique, **22**, 2, 195-300

SCHMERTMANN, J.H. (1966)
Discussion on Fletcher (1965)
Proc. ASCE, Jn. Soil Mech. Found. Div., SM5, 130-33

SCHMERTMANN, J.H. (1974)
Penetration testing in USA
Proc. 1st European Symp. on Penetration Testing (ESOPT 1), Vol. 1, 217-18

SCHMERTMANN, J.H. (1978)
Use the SPT to measure dynamic soil properties? – Yes, But...!
Dynamic Geotechnical Testing, ASTM SPT 645 pp. 341-55

SCHMERTMANN, J.H. (1979)
Statics of SPT
Jn. Geot. Div., Proc. ASCE, GT5, May, 655-70

SCHMERTMANN, J.H. and PALACIOS, A. (1979)
Energy dynamics of SPT
Jn. Geot. Engg. Div., Proc. ASCE, GTB, Vol. 105, pp 909-26

SCHNABEL, J.J. (1966)
Discussion on Fletcher, (1965)
Proc. ASCE, Jn. Soil Mech. Found. Div., **92**, SM2, 184

SCHULTZE, E. and MENZENBACH, E. (1961)
Standard penetration test and compressibility of soils
Proc. 5th Int. Conf. Soil Mechanics Foundation Engineering, Paris, Vol. 1, pp 527-32

SCHULTZE, E. and SHERIF, G. (1973)
Prediction of settlements from evaluated settlement observations for sands
Proc. 8th Int. Conf. Soil Mechanics Foundation Engineering, Moscow, Vol. 1, 3, 225-30

SEED, H.B., IDRISS, I.M. and ARANGO, I. (1983)
Evaluation of liquefaction potential using field performance data
Proc. ASCE, Jn. Geot. Engg, **109**, 3

SEED, H.B., TOKIMATSU, K., HARDER, L.F. and CHUNG, R.M. (1985)
Influence of SPT procedures in soil liquefaction resistance evaluations
Jn. Geot. Engg, Proc. ASCE, Dec., 1425-45

SEROTA, S. and LOWTHER, G. (1973)
SPT practice meets critical review
Ground Engg, **6**, 1, 20-3

SHIOI, Y., UTO, K., FUYUKI, M. and IWASAKI, T. (1981)
Standard Penetration Test
In: *Present State and Future Trend of Penetration Testing in Japan*, Paper III, 8-20. Jap. Soc. SMFE

SHIOI, Y. and FUKUI, J. (1982)
Application of N-value to design of foundations in Japan
Proc. 2nd Eur. Symp. on Penetration Testing (ESOPT II), Amsterdam, Vol. 1, 159-64

SIMONS, N.E. and MENZIES, B.K. (1977)
A short course in foundation engineering
IPC Science and Technology Press, London

SKEMPTON, A.W. (1986)
Standard Penetration Test procedures and the effects in sands of overburden pressure, relative density, particle size, ageing and over consolidation
Geotechnique, **36,** 3, 425-47

SKEMPTON, A.W. and BJERRUM, L. (1957)
A contribution to the settlement analysis of foundations on clay
Geotechnique, **7,** 4, 168-78

SMITH, E.A.L. (1962)
Pile-driving analysis by the wave equation
Trans. ASCE, **27,** 1, 1145-93

SOIL ASSOCIATION OF AUSTRALIA (SAA) Test 16A (1971)
Determination of Penetration Resistance of a Soil
Australian Standard Methods of Testing Soils for Engineering Purposes, Part VI – soil strength tests
Standards Association of Australia

STOKOE, K.H. and ABDEL-RAZZAK, K.G. (1975)
Shear moduli of two compacted fills
Proc. ASCE Conf. on In situ measurement of soil properties, Vol. 1, pp 422-49

STROUD, M.A. (1974)
The Standard Penetration Test in Insensitive Clays and Soft Rocks
Proc. Eur. Symp. on Penetration Testing (ESOPT I), pp 367-75

STROUD, M.A. (1989)
The Standard Penetration Test – its application and interpretation
Proc. ICE Conf. on Penetration Testing in the UK, Birmingham. Thomas Telford, London

STROUD, M.A. and BUTLER, F.G. (1975)
The Standard Penetration Test and the engineering properties of glacial materials
Proc. Symp. on Engineering Properties of Glacial Materials, Midlands Geotechnical Society, Birmingham, 117-28

STUBBINGS, J.E. (1966)
Discussion on Fletcher (1965)
Proc. ASCE, Jn. Soil Mech. Found. Div., **92,** SM2, 185-7

SUTHERLAND, H.B. (1963)
The use of in situ tests to estimate the allowable bearing pressure of cohesionless soils
The Structural Engineer, **41,** 3, 85-92

SYKORA, D.W. and STOKOE, K.H. (1983)
Correlations of in situ Measurements in Sands of Shear Wave Velocity, Soil Characteristics and Site Conditions
Geot. Engg. Rep. GR83-33, Dept. Div. Engg, University. of Texas at Austin

TS 1900 (1975)
Insaat muhendisliginde zemin deneyleri ('Methods of testing soils for civil engineering purposes')
Turk Standardler: Enstitusu, Ankara

TALBOT, J.C.S. (1981)
The prediction of settlements using in situ penetration test data
Unpublished MSc Dissertation, University of Surrey

TERZAGHI, K. (1947)
Recent trends in sub-soil exploration
Proc. 7th Texas Conf. Soil Mechanics and Foundation Engineering, 1-15. University Texas Bureau of Engineering Res. and Dept. of Civil Engineering, Austin, Spec. Pub. No. 17, 1-15

TERZAGHI, K. and PECK, R.B. (1948)
Soil Mechanics in Engineering Practice
John Wiley, New York (1st edn)

TERZAGHI, K. and PECK, R.B. (1967)
Soil Mechanics in Engineering Practice
John Wiley, New York (2nd edn)

THOMPSON, R.P. (1980)
Discussion on classification of weak rocks using the SPT
Proc. 7th Eur. Conf. Soil Mechanics Foundation Engineering, Brighton, Vol. 4, p 92

THOMPSON, R.P. and LEACH, B.A. (1989)
The application of the SPT in weak sandstone and mudstone rocks
Proc. ICE Conf. on Penetration Testing in the UK., Birmingham, 21-4. Thomas Telford, London

THOMPSON, R.P., NEWMAN, R.L. and DAVIS, P.D. (1990)
Advances in in situ testing of weak mudstone rocks
Proc. Conf. on the Engineering Geology of Weak Rocks, Leeds, Geological Society

THORBURN, S. (1963)
Tentative correction chart for the Standard Penetration Test in non-cohesive soils
Civ. Engg and Pub. Wks. Rev., June, 752-3

THORBURN, S. and MACVICAR, S.L. (1971)
Pile load tests to failure in the Clyde alluvium
Proc. ICE Conf. on Behaviour of Fills, 1-7, 53–4

THORBURN, S. (1986)
Field Testing: The Standard Penetration Test
In: *Site Investigation Practice: Assessing BS 5930*, (ed. Hawkins, A.B.) 21-6. The Geological Society, London

TOKIMATSU, K. (1988)
Penetration Testing for dynamic problems
Proc. ISOPT-1, **1**, 117-36

TOKIMATSU, K. and YOSHIMI, Y. (1983)
Empirical correlation of soil liquefaction based on SPT N-value and fines content
Soils and Foundations, **23**, 4 (Japan)

TOMLINSON, M.J. (1965)
Foundation design and construction
Pitman Publishing, London

TSUCHIYA, H. and TOYOOKA, Y. (1982)
Comparison between N-value and pressuremeter parameters
Proc. 2nd Eur. Symp. on Penetration Testing (ESOPT II), Amsterdam, pp 169-74

TWINE, D. and GROSE, W.J. (1990)
Discussion on overviews and field logging
Proc. Int. Chalk Symp., Brighton, 209-11. Thomas Telford, London

UNITED STATES BUREAU OF RECLAMATION (USBR) (1963)
Earth Manual
(1st edn, revised). Washington DC

UTO, K., FUYUKI, M. KONDO, H. and MORIHARA, M. (1973)
Experimental study on the mechanism of dynamic penetration of rod in the Standard Penetration Test' (in Japanese)
Proc. of the Faculty of Engineering, Tokai Univ. No. **1**, 73-94, 2, 55-65

UTO, K., FUYUKI, M., KONDO, H. and MORIHARA M. (1975)
Fundamental studies on the mechanism of dynamical penetration of rod from a viewpoint of wave theory
Proc. of the Faculty of Engineering, Tokai Univ. **II**, 9-30

VAN WEELE, A.F. (1989)
Prediction versus performance
Keynote speech,
Proc. 12th Int. Conf. on Soil Mechanics and Foundation Engineering, Rio de Janeiro, 4.

WAKELING, T.R.M. (1966)
Foundations on Chalk
Proc. Conf. on Chalk in Earthworks and Foundations, ICE, London, pp 15-23

WAKELING, T.R.M. (1970)
A comparison of the results of standard site investigation methods against results of a detailed geotechnical investigation in Middle Chalk at Mundford, Norfolk
Proc. Conf. on In-situ Investigations in Soils and Rocks, BGS, London, pp 17-22

WARD, W.H., BURLAND, J.B. and GALLOIS, R.W. (1968)
Geotechnical assessment of a site at Mundford, Norfolk, for a large proton accelerator
Geotechnique, **18**, 399-431

WEBB, D.L. (1970)
Settlement of structures on deep alluvial sand sediments in Durban, South Africa
Proc. BGS Conf. on In-situ investigations in Soil and Rock. BGS, London

WILLIAMS, B.P. and WAITE, D. (1993)
The design and construction of sheet-piled cofferdams
Special Publication 95, CIRIA, London

WOODLAND, A.R., NG, C.L. and CORKE, D. (1989)
High pressure dilatometer tests in Upper Chalk at Hull
Proc. ICE Conf. on Penetration Testing in UK, Birmingham, pp 153-57

WROTH, C.P. (1984)
The interpretation of in situ soil tests
24th Rankine Lecture. *Geotechnique*, **34**, 4, 447-92

WRIGHT, S.J. and REESE, L.C. (1979)
Design of large diameter bored piles
Ground Engg., Nov., 17-50

YAMASHITA, K., TOMONO, M., and KAKURAI, M. (1987)
A method of estimating immediate settlement of piles and pile groups
Soils and Foundations, **27**, 1, 61-76

YOSHIDA, Y., KOKUSHO, T. and MOTONORI, I. (1988)
Empirical formulas of SPT Blow-counts for gravelly soils
Proc. 1st Int. Symp. on Penetration Testing (ISOPT-I), Vol. 1

YOUD, T.D. (1973)
Factors controlling maximum and minimum densities of sands
Proc. Symp. on Evaluation of relative density and its role in geotechnical projects involving cohesionless soil. ASTM STP Vol. 523, pp 98-112

ZOLKOV, E. and WISEMAN, G. (1965)
Engineering properties of dune and beach sands and the influence of stress history
Proc. 6th Int. Conf. on Soil Mechanics and Foundation Engineering, Montreal, Vol. 1, pp 134-138

Appendix 1 International Reference Test Procedure IRTP

1 SCOPE

1.1 This specification describes the principles constituting acceptable test procedures for the SPT from which results are comparable.

1.2 The SPT determines the resistance of soils in a borehole to the penetration of a tubular steel sampler, and obtains a disturbed sample for identification. The penetration resistance can be related to soil characteristics and variability.

The basis of the test consists of dropping a hammer weighing 63.5 kg onto a drive head from a height of 760 mm. The number of blows (N) necessary to achieve a penetration by the steel tube of 300 mm (after its penetration under gravity and below a seating drive) is regarded as the penetration resistance (N).

2 BORING METHODS AND EQUIPMENT

2.1 Boring methods

2.1.1 The boring equipment shall be capable of providing a reasonably clean hole to ensure that the penetration test is performed on relatively undisturbed soil.

2.1.2 When wash boring, a side-discharge bit should be used and not a bottom-discharge bit. The process of jetting through an open tube sampler and then testing when the desired depth is reached shall not be permitted.

2.1.3 When shell and auger boring with temporary casing, the drilling tools shall have diameters not more than 90% of the internal diameter of the casing.

2.1.4 When boring in soil that will not allow a hole to remain stable, casing and/or mud shall be used.

2.1.5 The diameter of the borehole should be between 63.5 and 150 mm.

2.2 Tubular steel sampler assembly

The tube of the sampler shall be made of hardened steel with a smooth surface externally and internally. The external diameter shall be 51 mm plus or minus 1 mm and the internal diameter throughout shall be 35 mm plus or minus 1 mm. Its length shall be 457 mm minimum.

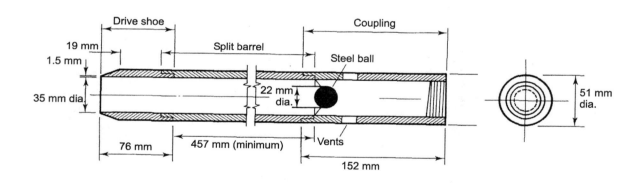

Figure 52 Cross section of SPT Sampler

The lower end of the tube shall have a driving shoe 76 mm long plus or minus 1 mm having the same bore and external diameter as the tube. Over the lowermost 19 mm it will taper uniformly inwards to reach the bore at the bottom edge. The material shall be the same as the tube. The drive shoe shall be replaced when it becomes damaged or distorted.

At the upper end of the tube a steel coupling shall be fitted to connect with the drive rods. Inside shall be a non-return valve with wide vents in the coupling wall, which are of sufficient size, to permit unimpeded escape of air or water on entry of the sample. The valve should provided a watertight seal when withdrawing the tubular steel sampler.

On acceptable form of the sampler assembly is shown in Figure 52.

2.3 Steel drive rods

2.3.1 The steel drive rods, connecting the sampler assembly to the hammer assembly shall have a section modulus appropriate to their total length and lateral restraint.

Appropriate section properties are:

Rod diameter (mm)	Section modulus ($\times 10^{-6}$ m^3)	Rod weight (kg/m)
40.5	4.28	4.33
50	8.59	7.23
60	12.95	10.03

Rods heavier than 10.03 kg/m shall not be used.

2.3.2 Only straight rods shall be used and periodic checks shall be made on site. When measured over the whole length of each rod the relative deflection shall not be greater than 1 in 1000.

2.3.3 The rods should be tightly coupled by screw joints.

2.4 Hammer assembly

2.4.1 The hammer assembly shall comprise:

   a) A steel drive head tightly screwed to the top of the drive rods. The energy transferred on impact shall be maximised by a suitable design of drive head.

   b) A steel hammer of 63.5 kg (plus or minus 0.5 kg) weight.

   c) A release mechanism which will ensure that the hammer has a free fall of 760 mm.

2.4.2 The guide arrangement shall permit the hammer to drop with minimal resistance.

2.4.3 The overall weight of the hammer assembly shall not exceed 115 kg.

2.4.4 In situations where comparisons of SPT results are important calibrations will be made to evaluate the efficiency of the equipment in terms of energy transfer.

3 TEST PROCEDURE

3.1 Preparing the borehole

3.1.1 The borehole shall be carefully cleaned out to the test elevation using equipment that will ensure the soil to be tested is not disturbed.

3.1.2 When boring below the groundwater table or in sub-artesian conditions the water or mud level in the borehole shall at all times be maintained at a sufficient distance above the groundwater level to minimise disturbance. The water or mud level in the borehole shall be maintained to ensure hydraulic balance at the test elevation.

3.1.3 The drilling tools shall be withdrawn slowly to prevent loosening of the soil to be tested.

3.1.4 When casing is used, it shall not be driven below the level at which the test is to commence.

3.2 Executing the test

3.2.1 The sampler assembly shall be lowered to the bottom of the borehole on the drive rods with the hammer assembly on top. The initial penetration under this total deadweight shall be recorded. Where this penetration exceeds 450 mm the test drive will be omitted and the 'N' value taken as zero.

After the initial penetration, the test will be executed in two stages:

Seating drive: A penetration of 150 mm. If the 150 mm penetration cannot be achieved in 50 blows, the latter shall be taken as the seating drive.

Test drive: A further penetration of 300mm. The number of blows required for this 300 mm penetration is termed the *penetration resistance* (N). If the 300 mm penetration cannot be achieved in 100 blows the test drive shall be terminated.

The rate of application of hammer blows should not be excessive such that there is the possibility of not achieving the standard drop or preventing equilibrium conditions prevailing between successive blows. Typically, the maximum rate of application of blows is 30 per minute. The number of blows required to effect each 150 mm of penetration shall be recorded. If the seating or test drive is terminated before the full penetration, the record should state the depth of penetration for the corresponding 50 blows.

3.3 Recovery of soil sample and labelling

3.3.1 The sampler shall be raised to the surface and opened. The representative sample or samples of the soil in the sampler shall be placed in an air-tight container.

3.3.2 Labels shall be fixed to the containers with the following information:

    a) Site

    b) Borehole number

    c) Sample number

    d) Depth of penetration

    e) Length of recovery

    f) Date of sampling

    g) Standard penetration resistance (N).

4 REPORTING OF RESULTS

The following information shall be reported:

1. Site
2. Date of boring to test elevation
3. Date and time of commencement and end of test
4. Borehole number
5. Boring method and dimensions of temporary casing, if used
6. Dimensions and weight of drive rods used for the penetration tests
7. Type of hammer and release mechanism or method
8. Height of free fall
9. Depth to bottom of borehole (before test)
10. Depth to base of casing
11. Information on the groundwater level and the water or mud level in the borehole at the start of each test
12. The depth of initial penetration and the depths between which the penetration resistances (seating and test drives) were measured
13. Penetration resistances (seating and test drives)
14. The descriptions of soils as identified from the samples in the sampler
15. Observations concerning the stability of strata tested or obstructions encountered during the tests etc., which will assist the interpretation of the test results
16. Calibration results, where appropriate.

EXPLANATORY NOTE

Calibrations of drive rods and hammer assemblies, where appropriate, would normally be carried out for each rig and separately from the investigation work. They would be applicable to a particular project, based upon the personnel and equipment involved.

Appendix 2 SPT energy measurement

A2.1 INTRODUCTION

Energy measurement is particularly important when SPT hammers with unknown characteristics are in use. In the UK very limited measurements have been made, at the time of writing. But as this Appendix shows, it is likely that UK automatic trip hammers, when properly maintained and used sensibly, will give a high standard of uniformity to the energy delivered by their blows. For British engineers, then, the effects of highly variable and unknown levels of SPT energy are likely to be most significant when carrying out site investigations abroad, or using the SPT data obtained in other countries.

This Appendix gives information on how SPT hammer energy may be measured. ASTM Standard D4633-86 may be used to specify energy measurement in contract documents, since at the time of writing there is no British Standard for this activity.

A2.2 THE SIGNIFICANCE OF ENERGY TRANSMISSION

Palmer and Stuart (1957) first suggested, and Schmertmann and Palacios (1979) have shown experimentally that, at least up to $N = 50$, penetration resistance varies inversely with the energy, E, transmitted down the drill rods to the split spoon. Therefore

$$\frac{N_1}{N_2} = \frac{E_1}{E_2}$$

and through this equation N values may be compared if the energies in both tests are known. Alternatively, as has been suggested by Seed et al. (1985) and subsequently Skempton (1986), SPT results may be corrected to give values equivalent to some standard energy input.

The free-fall energy delivered by a 63.5 kg (140 lb) mass falling through 760 mm (2 ft 6 in) is 473.4 Joules. But due to frictional losses on catheads, in sheave bearings and on trip hammer shafts this is not entirely converted into kinetic energy as the weight falls. In some tests the weight may not be dropped through the correct height and it is thought that further reductions may occur as a result of energy spent in accelerating the mass of a heavy striker plate (anvil) and as a result of bending very light drill rods.

The energy transmitted down the SPT rods can conveniently be expressed in terms of the Rod Energy Ratio, which is a measure of the efficiency of a given hammer system

$$ER_r = \frac{100 E_r}{E^*}$$

where ER_r = rod energy ratio (%)
E_r = measured energy passing down rods
E^* = theoretical free-fall energy of standard hammer.

Two factors are important in considering hammer efficiency; the average efficiency and the blow-by-blow variability. Both of these vary with hammer design, although there are fewer published data on blow-by-blow variability, which can be particularly high for some hammer types. For example, for a single American slop-rope hammer, Kovacs et al. (1977) reported blow-by-blow rod energy ratios which varied from as little as 35% to as much as 69%. As the summary in Table 19 shows, average rod energy ratios for different hammers vary only slightly more.

Table 19 *Variation of rod energy ratios for SPT hammers*

Country	Hammer Type	Release mechanism	Average rod energy ratio (%)	Source
Argentina	donut	cathead	45	1
Brazil	pin weight	hand dropped	72	3
China	automatic donut	trip hand	60	1
	donut	dropped cathead	55	2
			50	1
Colombia	donut	cathead	50	3
Japan	donut	Tombi trigger	78-85	1, 4
	donut	cathead-2 turns + special release	65, 67	1, 2
UK	authomatic	trip	73	5
USA	safety	2 turns on cathead	55-60	1, 2
	donut	2 turns on cathead	45	1
Venezuela	donut	cathead	43	3

References: 1. Seed et al. (1985) 4. Riggs (1986)
2. Skempton (1986) 5. This Appendix
3. Decourt (1990)

A2.3 HAMMER DESIGN

Hammer mechanisms can be categorised as:

Automatic trip hammers. Here the height of fall is controlled by the hammer mechanism, and the weight is dropped freely once it is lifted to the correct height. Energy may be lost in shaft friction (particularly if the hammer is not held vertically) and in momentum effects due to the mass of the anvil.

Hand-controlled trip hammers. The height of fall is estimated for each blow by the driller, but a trigger mechanism is used to ensure no energy loss during the weight fall.

Slip-rope hammers. The weight is lifted using a continuously rotating cathead, and once it reaches an estimated height of 760 mm the rope is loosened. While the weight falls, energy is lost in friction not only by the rope on the cathead drum, but also in the sheave bearing on the drilling rig mast.

Hand-lifted hammers. In this, the earliest hammer mechanism, the weight is lifted by two men pulling down on a rope which passes over a sheave at the top of a tripod placed over the borehole. Even when released carefully, energy will be lost in the sheave bearing.

Apart from frictional losses, either in sheaves or on catheads, the weight of the anvil (or striker plate, or knocking head) is thought to influence the energy delivered to the rods. Small anvils, such as those used in Japan and with pin weight hammers in Brazil, weigh only about 2 kg. In contrast, British trip hammer anvils weigh as much as 20 kg.

A2.4 THEORY

A2.4.1 Force-time equation

When one end of a long elastic rod is displaced at a constant velocity, v_o, for a time dt, a stress with amplitude σ_o propagates a distance dx from the end of the rod, which displaces du.

Thus

$$dx = c.dt$$

$$du = v_o.dt \qquad \ldots (1)$$

where c is the propagation velocity of the stress wave.

The strain induced by stress, σ_o, over length dx is:

$$\frac{\partial u}{\partial x} = \frac{v_0}{c} = \frac{1}{c}.\frac{\partial u}{\partial t} \qquad \ldots (2)$$

Therefore, from equation (2), and relating stress to strain through Young's Modulus, E,

$$\sigma_0 = E.\frac{\partial u}{\partial x} = \frac{E.v_0}{c} \qquad \ldots (3)$$

Equation (3) was derived by Thomas Young in 1807. It can be expressed in a number of forms:

$$\sigma_0 = \frac{E.v_0}{c} = \sqrt{\rho E.v_0} = \rho c v_0 = \sqrt{\frac{\gamma E}{g}}.v_0 = \frac{\gamma c}{g}.v_0 \qquad \ldots (4)$$

where γ is the weight/unit volume of the rod
  g is the gravitational constant
  ρ is the density of the rod.

If, instead of being displaced at a constant velocity, the end of the rod is struck by a rigid body, mass M, travelling at an initial velocity v_o, the initial stress in the rod, σ_o, at the moment of impact is:

$$\sigma_0 = \frac{\gamma c v_0}{g}$$

During impact the velocity of the rigid body, mass M, will decrease due to the force imposed on it by the rod. As its velocity decreases, so also does the stress on the impact surface. For force equilibrium, the force causing deceleration must equal the force at the top of the rod, i.e. for a rod with cross-sectional area A

$$M\frac{dv}{dt} + A\sigma = 0 \qquad \ldots (5)$$

Substituting Young's relationship for σ (equation (4)) into equation (5) yields

$$\frac{dv}{dt} + \frac{\gamma A}{Mg}.cv = 0 \qquad \ldots (6)$$

Integrating with respect to t, and setting $v = v_o$ at $t = 0$, leads to

$$v = v_0.\exp\left(-\frac{\gamma A}{Mg}.ct\right) \qquad \ldots (7)$$

Substituting for σ and σ_o (equation (4)) gives

$$\sigma = \sigma_0.\exp\left(-\frac{\gamma A}{Mg}.ct\right) = \sigma_0.\exp\left(-\frac{mct}{L}\right) \qquad \ldots (8)$$

where $m = \dfrac{AL\gamma}{Mg}$

L is the length of the rod.

This expression was derived by St. Venant in 1883. It is implicit that the rod is either long, or that its far end is fixed, so that whole-body translation is prevented.

Equation (8) can be re-written, introducing $F = \sigma.A$, to give

$$F = F_0.\exp\left(-\frac{mct}{L}\right) = F_0.\exp\left(-\frac{A\gamma ct}{Mg}\right) \qquad \ldots (9)$$

where, from equation (4),

$$F_0 = \frac{AEv_0}{c}$$

The theoretical force-time relationship for a 63.5 kg (140 lb) hammer weight falling 762 mm (2ft 6in) on to the end of an infinitely long AW rod is shown in Figure 53.

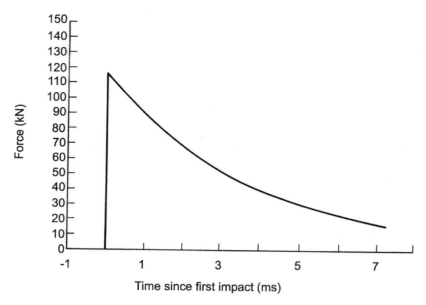

Figure 53 *Theoretical force-time relationship for a 63.5 kg hammer striking the end of an infinitely long AW rod*

A2.4.2 Energy equation

Now consider the incremental energy contained within the rod length dx stressed to σ_o at the start of compression. The total energy, E_t, has components of strain energy, E_e, and kinetic energy, E_k. The incremental energy components are

$$dE_0 = \frac{\sigma_0.A.du}{2} = \frac{\sigma_0^2.A.c.dt}{2E} = \frac{\sigma_0^2.A.dx}{2E} \qquad \ldots (10)$$

$$dE_k = \frac{A.\gamma.dx}{2g}v_0^2 = \frac{\sigma_0^2.c^2.\gamma.A.dx}{2E^2 g} = \frac{\sigma_0^2.A.dx}{2E} \qquad \ldots (11)$$

Thus $dE_e = dE_k$, and the total rod energy consists of equal parts of strain energy and kinetic energy. Since it is easier to measure strain energy, the total energy in the entire rod length may be obtained from:

$$E_t = 2E_e = 2\int_0^x \frac{\sigma x^2.A}{2E}.dx = 2\int_0^t \frac{\sigma t^2.A.c}{2E}.dt$$

$$E_t = \frac{c}{A.E}\int_0^t F(t)^2.dt \qquad \ldots (12)$$

The actual energy delivered by a real hammer can be determined by insetting a force transducer in the AW rod strong, measuring the force-time relationship for a number of blows, and using equation 12 to determine the total energy, E_t.

A2.4.3 Short-rod correction for measured energy

The theoretical total energy derived by force measurement can be obtained by combining equations (9) and (12) to give

$$E_t = \frac{c}{A.E} \int_0^t F_0^2 \exp\left(-\frac{2.A.\gamma.c.t}{Mg}\right) dt \qquad \ldots (13)$$

In practice it is not always possible to use a rod string long enough to be considered infinite. Once the stress wave reaches the base of the rods it will be reflected, as a compressive wave if the SPT split spoon is in a strong material, or as a tensile wave if the soil is weak.

Figure 54 shows the typical layout of hammer, rods, split spoon and load cell for energy measurement. The load cell should be placed no closer than 10 rod diameters below the underside of the anvil, to avoid effects due to stress concentrations at the top of the rods. Since the reflected wave travels down the rod and back up as a velocity c (normally 5.1 metres/millisecond), it arrives at the load cell at

$$t' = 2.\frac{(L-l)}{c} \qquad \ldots (14)$$

after the first part of the wave reaches the load cell when travelling in a downward direction. After this time (i.e. t') the measured force-time relationship cannot be used in equation 12, because it contains elements of up-going energy.

The maximum energy that could be measured during the period $t = 0$ to $t = t'$ for a theoretical force-time relationship would be

$$E_t' = \frac{c}{A.E} \int_0^{t'} F_0^2 \exp\left(-\frac{2.A.\gamma.c.t}{Mg}\right).dt \qquad \ldots (15)$$

A correction, K, must be applied to the measured energy, and this is based on the theoretical force-time relationship:

$$K = \frac{E_t}{E_t'} \qquad \ldots (16)$$

From equation (13)

$$E_t = \frac{c}{A.E} F_0^2 \left[-\frac{L}{2.m.c} \exp\left(-\frac{2.m.c.t}{L}\right)\right]_0^t$$

$$E_t = \frac{c}{A.E} F_0^2 \frac{L}{2.m.c} \left[1 - \exp\left(-\frac{2.m.c.t}{L}\right)\right] \qquad \ldots (17)$$

Since

$$m = \frac{A.L.\gamma}{M.g}$$

and

$$F_0^2 = \left(\frac{\gamma.E}{g}\right)v_0^2.A^2$$

$$E_t = \frac{Mv_0^2}{2}\left[1 - \exp\left(-\frac{2.m.c.t}{L}\right)\right]$$

$$E_t = M.g.h\left[1 - \exp\left(-\frac{2.m.c.t}{L}\right)\right] \qquad \ldots (18)$$

where h is the free height of fall of the rigid body, the hammer, mass M.

Substituting

$$t = t' = 2.\frac{(L - l)}{c}$$

in equation (18) yields

$$E_t' = M.g.h\left[1 - \exp\left(-4.m.\left\{1 - \frac{l}{L}\right\}\right)\right] \qquad \ldots (19)$$

while substituting $t = \infty$ (for an infinitely long rod) yields

$$E_t = M.g.h \qquad \ldots (20)$$

Dividing equation (19) by equation (20) yields the correction factor, K, i.e.

$$K = \frac{1}{\left[1 - \exp\left(-\frac{4.A.\gamma.L}{Mg}\left\{1 - \frac{l}{L}\right\}\right)\right]}$$

For standard AW rods, this correction is shown in Figure 55. It can be seen that a rod length of 15 m is necessary to ensure that the measured energy is within 2% of the actual value.

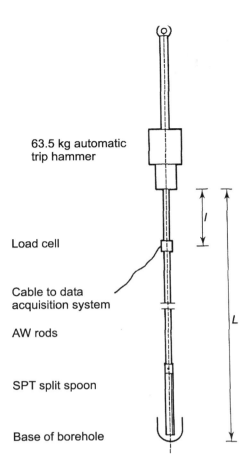

Figure 54 *Layout of equipment during energy measurement*

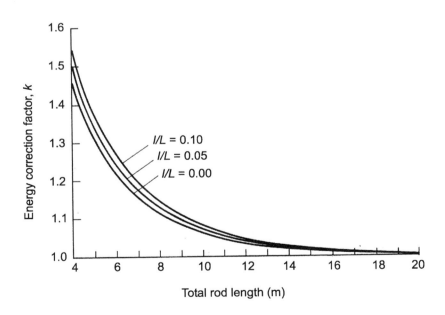

Figure 55 *SPT energy correction for short rods and load cell position*

A2.5 MEASURING TECHNIQUES AND RESULTS OF MEASUREMENTS

Three techniques have been used, and are described below (Figure 56). In two sets of tests a commercial load cell was used, whilst in the third a standard AW rod was strain-gauged. In two tests, one involving a load cell and the other the strain-gauged AW rod, the output was passed through amplifiers and recorded using a transient recorder, and then down-loaded to a desk-top computer for processing. In the third a commercially available geotechnical logging system with a purpose-built SPT measuring module was used. From these tests the following experience was gained.

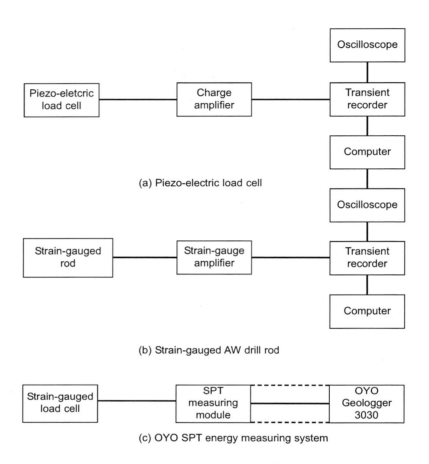

Figure 56 *SPT energy measuring systems*

1. When load cells are used they should be calibrated when connected in the rod in which they will be used, to ensure that changes in cross-sectional area in the vicinity of the load cell do not give false readings. These problems can be minimised by ensuring that the rod dimensions do not change for approximately 10 rod diameters above and below the load cell.

2. Strain gauge load cells perform satisfactorily, as will piezo-electric load cells. However, although piezo-electric load cells have a fast response which makes them suitable for dynamic measurements of this type, they require the use of a charge amplifier before conventional amplication can take place. This additional complication is avoided when using conventional strain-gauged load cells.

3. Amplifiers must be sufficiently fast. The ability of an amplifier to perform at high frequency can be assessed on the basis of its unity band width. If too slow an amplifier is used then higher frequencies are attenuated, and a low energy may be measured.

4. Battery-powered equipment overcomes the significant problems of mains-borne noise which can occur when mains-driven equipment is used, and also gives superior results to those obtained when a portable generator is used to provide power.

As might be expected, the most convenient equipment to use was a commercially available SPT measuring module (the Oyo Geologger 3030 shown in Figure 57).

Field testing was carried out on five occasions, using a Duke and Ockenden (DANDO) automatic trip hammer, details of which are given in Figure 8. Table 20 gives the results of the measurements made in those tests which were considered reliable, due allowance being made for the various factors discussed above. AW rods and a standard split spoon (BS 1377: 1975) were used throughout, in a 100 mm diameter borehole. In the first tests, using a piezo-electric load cell, high energy levels were measured. Once the short rod correction was applied the rod energy levels rose above 100%. Since the load cell could not be calibrated under static conditions, this measuring system was abandoned. In all subsequent tests, including those carried out using the Oyo Geologger 3030 SPT measuring module, a 1.5 m rod containing the load cell or a strain-gauged central section was calibrated in a Satec 500 kN Class A universal testing machine.

The results of tests given in Table 20 show good agreement between the average corrected rod energy ratios in the last four sets of measurements, and indicate a standard deviation averaging only 2.7%. The 95% confidence limits are therefore +/- 5.4% of the average rod energy ratio. In view of the small standard deviations of rod energy, the differences between the last two measurements are significant.

Figures 58 (a) and (b) show typical force-time relationships from each of these series of tests; it is difficult to see differences which might explain the variation in measured energy. Shown on Figure 58 is the time, t′, at which it is predicted (on the basis of wave velocity of 5.1 m/ms) that the reflected event will appear on the trace. Integration of the square of the wave-form is carried out up to this point during the calculation of rod energy. It can be seen that a rapid drop in force occurs after this point, as a tensile wave passes over the load cell from below. The equivalent penetration resistance for the upper trace was 30 blows/300 mm, whilst for the lower trace it was 49 blows/300 mm.

Thus measurements of SPT rod energy for a single British automatic trip hammer have shown that this hammer delivers energy in a highly repeatable way, as demonstrated by the very low standard deviations calculated from over 120 blows. For tests 2, 3 and 4 (Table 20) the average corrected rod energy ratio is 73%, with a standard deviation of 2.8%.

It was suggested by Skempton (1986) that British automatic trip hammers deliver about 60% of the available free-fall energy, but the tests reported here show that 75% is a more appropriate figure. In the final test a lower average of 69% was obtained, but it is suggested that this value should be regarded as less reliable, firstly because the penetration resistance was high and secondly because the test was carried out in an unsupported hole where the rods had been inserted on the previous day and left overnight. It is possible that the sides of the lower part of the borehole had collapsed, providing some shear resistance to the lower part of the rods.

Of the three measurement systems used in this study, the purpose-built Oyo Geologger SPT measuring module was found to be the most convenient and reliable in use. The system is battery powered, has an internal printer, and stores data on a 3½ in disk. The two systems using a transient recorder both performed reasonably well, but because of the need to provide 240V AC for the computer, oscilloscope and transient recorder, some noise problems were encountered, and were particularly bad when using mains supplies. The frequency response of the strain-gauge amplifiers did not prove to be a problem.

Figure 57   *Energy measurement using the Oyo Geologger 3030 SPT measuring module*

Table 20 *SPT energy measurements on a DANDO automatic trip hammer*

Test No.	Method of measurement	Rod length, L (m)	Distance to load cell, l (m)	No. of blows	Measured rod energy ratio (average) (%)	SD (%)	Corrected rod energy ratio (%)
1	Piezo-electric load cell + transient recorder	8.25	0.75	8	92.7	3.5	107.5
2	Strain-gauged AW rod, slow amplifier + transient recorder	8.25	0.75	28	65.0	3.8	73.5
3	Strain-gauged AW rod, fast amplifier + transient recorder	12.75	0.75	6	70.6	4.4	73.0
4	Oyo Geologger 3030 + SPT measuring module	13.31	0.86	44	72.8	1.9	74.6
5	Oyo Geologger 3030 + SPT measuring module	13.31	0.86	50	67.5	2.7	69.2

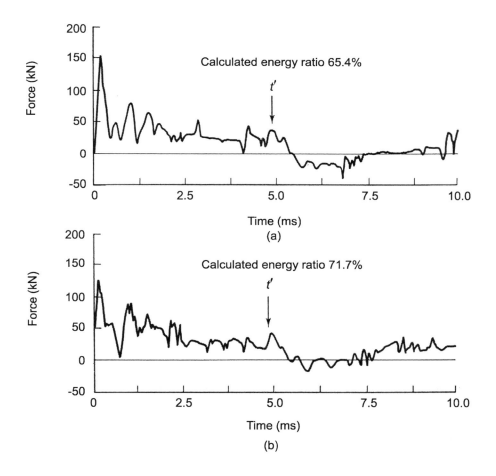

Figure 58 *Typical force-time relationships from SPT blows*